FBI

DIGITAL EXHAUST
OPT OUT GUIDE

HOW TO MAKE YOURSELF A HARD TARGET ONLINE

1 TABLE OF CONTENTS

2 DISCLAIMER

2.1 PURPOSE
The Digital Exhaust Opt Out Guide 2.0 supersedes version 1.0, which was published in October 2019 and is being updated as of October 2021. This Guide was created to mitigate risk for Law Enforcement employees' and their families as it pertains to protecting their personal information, which is vulnerable to exploitation. This risk includes potential for threat actors to identify, target, and track anyone affiliated with Law Enforcement via use of open source, Internet-based services offering searches of data aggregated about the American public. To mitigate this risk, this Guide was created as a first-of-its-kind aid for the Law Enforcement community in highlighting and presenting recommendations to reduce these vulnerabilities. This document is for informational purposes only. Questions about this document can be directed to the email address listed below in Section 2.5.

2.2 LIMITING LIABILITY
This Digital Exhaust Opt Out Guide was prepared as a collection of best practices to assist Law Enforcement employees. Neither the United States Government nor any agency thereof, nor any of their employees, makes any warranty, express or implied, or assumes any legal liability or responsibility for the accuracy, completeness, or usefulness of any information, or process disclosed. Reference herein to any specific commercial product, process, or service by trade name, trademark, manufacturer, or otherwise does not necessarily constitute or imply its endorsement, recommendation, or favoring by the United States Government or any agency thereof. The views and opinions of authors expressed herein do not necessarily state or reflect those of the United States Government or any agency thereof.

2.3 LINKS
The appearances of hyperlinks, which are external to Law Enforcement databases, are provided as a convenience and for informational purposes only; they do not constitute endorsement by the Federal Bureau of Investigation. The Federal Bureau of Investigation bears no responsibility for the accuracy, legality or content of the external site or for subsequent links. Contact the external site for answers to questions regarding its content. The links provided within this Guide are current as of the publication in October 2019.

2.4 CONTENT
No policy may contradict, alter or otherwise modify the standards of your Law Enforcement agency. Nothing in this Guide supersedes existing law and/or Department of Justice policy. Precautions must be taken to ensure this information is stored and/or destroyed in a manner that precludes unauthorized access.

2.5 AVAILABILITY

If you have questions, concerns or comments regarding the Digital Exhaust Opt Out Guide, please direct any inquiries to the email address kc_digitalexhaust@fbi.gov.

3 WHAT IS DIGITAL EXHAUST?

Digital Exhaust is data on the Internet about you.[1][2] It is all the information or 'consumer data' a person creates as they interact with web sites and services. You create some of it and others create some of it about you.[3,4] These data points are exploitable to find, target, and track you.[5] Your Digital Exhaust holds extremely sensitive information that identifies you and reveals your private activities. Controlling Digital Exhaust is possible but complex.[6] This document serves to make it easy, or at least easier.

3.1 WHY SHOULD YOU CARE?

Because *your* privacy matters. Consider the vast amounts of personal information that different services hold about us and be mindful of what you give other organizations access to.[7] The privacy choices you make can have lasting impacts on you and your loved ones for better or worse.[8] This guide is laid out for you in a way that is the key difference in aiding FBI employees and their families in opting out of their data and taking positive steps towards keeping their Digital Exhaust from repopulating and out of the hands of a variety of threat actors.[9]

3.2 WHY DO I NEED A GUIDE?

Every interaction you have with the internet and technological tools leaves a trace, and these traces can be valuable.[10] Heading into this blindly will consume and waste a lot of your time.[11] Not anymore. These preventative measures are simple enough to employ and use safely in everyday life, both physically and online, while comprehensive enough to deny spectrum access to threat actors who could gain important operational advantages at the expense of you – an FBI employee – or your family.[12]

3.3 WHERE DO I ACTUALLY FIT INTO THE DIGITAL EXHAUST LANDSCAPE?

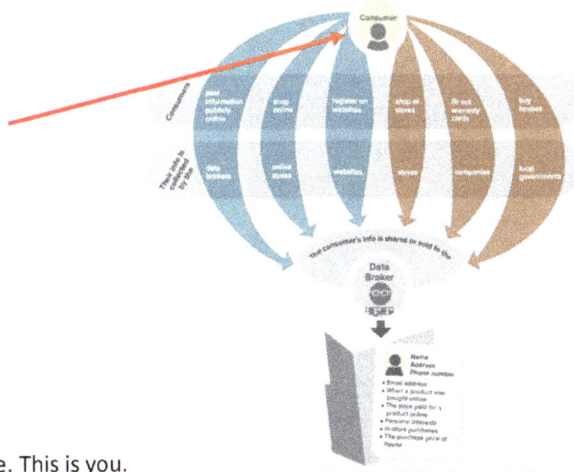

Here. This is you.

Figure 1. Digital Exhaust Ecosystem.

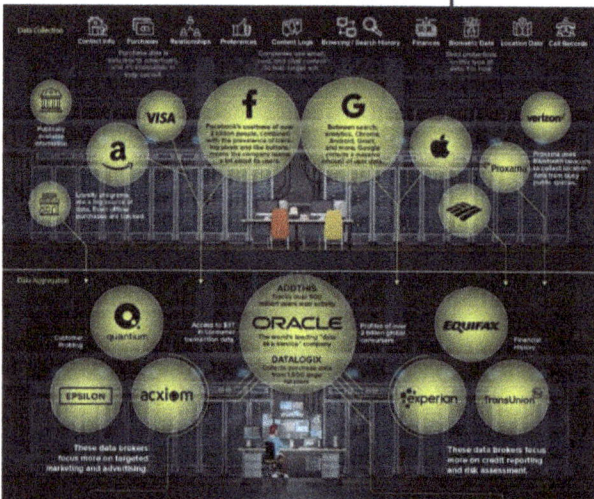

This is also you.

Figure 2. Digital Exhaust Ecosystem Players[13]

4   TOPS FRAMEWORK

To combat Digital Exhaust, it is recommended that users conduct a personal risk assessment of what they define as acceptable levels of risk for themselves and their family.

- This personal risk assessment often involves users assessing what pieces of their personal information form key assets, what they can remove online, what they cannot remove online, what they can obfuscate through deception and/or disinformation or simply allowing errors that may exist with Data Brokers and Data Aggregation websites to hold misinformation which also obfuscates an identity or exact personal information.
- Before a user can conduct a risk assessment, it is important they have the right mindset and then use a framework. One framework they can use is called **TOPS**.
- TOPS stands for **Threats, Opportunities, Preventative Measures** and **Strengths**. This framework is applied as follows:

**Digital Exhaust
TOPS Framework**

Threats	Opportunities	Preventative Measures	Strengths
• Who are the threat actors I am most concerned with researching my Digital Exhaust?	• What opportunities does my Digital Exhaust allow for exploitation by assessed threat actors?	• What are preventative measures I can take to address weaknesses in my Digital Exhaust?	• What are my strengths in regards to where my Digital Exhaust is controlled?

4.1 THREATS

Using **TOPS** helps a user *SPOT* who they assess to be their biggest Threat Actors and prioritize where they invest their time to minimize the impact on their life.[14]

- *"Who are the threat actors I am most concerned with researching my Digital Exhaust?"*[15]

4.2 OPPORTUNITIES

Using **TOPS** always reminds a user of what they **POST** online.

- *"What opportunities does my Digital Exhaust allow for exploitation by who I assess to be my Threat Actors?"[16]*

4.3 PREVENTATIVE MEASURES

Using **TOPS** helps a user **STOP** problems through mitigation.

- *"What are preventative measures I can take to address my weaknesses in my Digital Exhaust?"[17]*

4.4 STRENGTHS

Using **TOPS** helps a user's decision-making as it pertains to what *OPTS* I choose to execute.

- Do I opt out of data?[18]
- Do I opt into a service to help me control my personal information? [19]
- Do I opt to create disinformation, which pollutes the data broker ecosystem? [20]
- Do I opt to do nothing and allow misinformation to circulate to my advantage? [21]

It all factors into the question a user can ask through **TOPS**:

- *"What are my strengths in regard to where my Digital Exhaust is controlled?"[22]*

4.4.1 TOPS Output

As it pertains to user's Digital Exhaust, you can use this framework and choose what makes up your personal information's Key Assets.

- It is only then a user can begin assessing how their Digital Exhaust can be exploited and can begin building preventative or protective measures to mitigate their risk across a spectrum of tracking capabilities their family and they face personally.
- This framework can aid a user in understanding and shifting how they interact with their Web Browser, Mobile Phone and Mobile Apps, Social Media platforms and the totality of their privacy settings, which can be used for their benefit. The primary issue for a user is and will always be the intent of who can exploit the totality of their Digital Exhaust and for what purpose.[23]

4.5 PERSONAL INFORMATION "KEY ASSETS"

Personal Information key assets are critical pieces of a user's personal information that deserve special protection because of their destructive potential.[24]

- This Guide defines destructive potential as any key assets that if exposed publicly, could help targeting efforts by threat actors who could endanger a user's family or themself through intimidation or physical violence and/or damage my reputation or finances through identity theft or financial swindles.

4.5.1   Types of Key Assets

How does a user show what key assets within their personal information require special protection?

- For this Guide, the following key assets are the ones that users should apply preventative measures to include:

- *First and Last Name*
- *Date of Birth*
- *Home Address*
- *Social Security Number*
- *Username(s)*
- *E-Mail Address(es)*
- *IP Address(es)*
- *Telephone Number(s)*
- *Credit Card Number(s)*

4.5.2   Preventative Measures Applied To "Key Assets"

Once a user has named what threat actors they may meet, they can begin evaluating the totality of preventive measures and tailor them to be employed to thwart specific or all threat actors. These preventative measures may range from:

- Ensuring simple privacy settings are configured correctly.
- To mitigating physically consequential risks associated with their personal telephone number, home address and people search sites.
- Mitigating advanced threats such as ensuring a user has properly reduced any emissions of their Digital Exhaust on issues such as Online Behavioral Advertising, which looks to use a user's Activity-Based Intelligence to figure out their Patterns-of-Life, through Mobile Advertising, Behavioral Targeting, Categorical Targeting, Retargeting, Search Retargeting, and Dynamic Ads.
- To more subtle yet intrusive issues like ensuring a user has
  - Mitigated Intelligent Tracking Prevention techniques,
  - Identified and disabled location tracking,
  - Disabled their photo's metadata,
  - Ensured they have deidentified their debit and credit card's ability to track their card transaction data,
  - And prevented their Web Browser from actively exploiting their Browser's unique fingerprint.

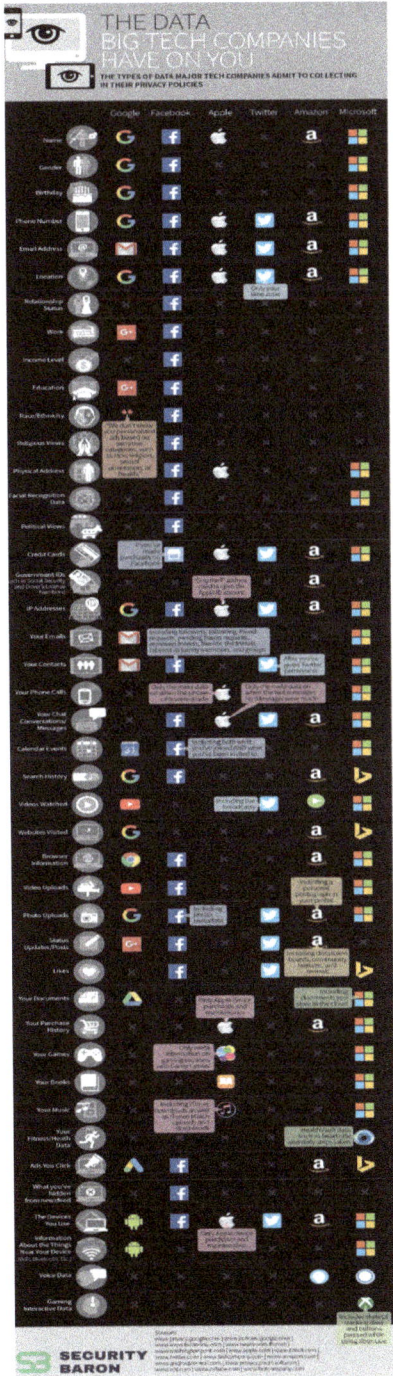

5 THE GUIDE

Presently, the FBI actively investigates a broad range of threat actors, many of whom have resources and technical abilities that can be used to target FBI employees and their families.

- These threat actors will continue to exploit the ever-increasing variety, volume, and speed of data sources to target FBI employees and their families, which requires the deployment of preventative measures.
- Perform these opt out steps to control your digital exhaust. Progress through the Guide in the order presented for best results.

5.1 SECURING YOUR WEB BROWSER

Tracking of browsing behavior is part of the daily routine of internet use.[25] Companies use it to adapt ads to the personal needs of potential clients or to measure their range.[26]

5.1.1 Tracking Cookies

Cookies are a way to store user settings for websites locally in the browser.[27] For example, you might set your preferred time zone, which would result in a cookie being created in your browser with that setting.

5.1.2 Browser Fingerprinting

Browser Fingerprinting, which is difficult to block, is based on the idea that every computer configuration is unique in some way.[28]

- A lot of that data is directly available to the sites you visit, usually for compatibility purposes.
- While cookie tracking works by placing a unique identifier on a person's web browser, fingerprinting takes place when a company creates a profile of your device's unique characteristics.

All web browsers collect the following 10 types of data about you:
1. Your hardware and software.
2. Your connection information (to include your IP address and browser speed).
3. Your geolocation data.
4. Your browsing history.[29]

TLP:GREEN

5. Your mouse or touch pad movements.
6. Your device's orientation (if using a Mobile browser).
7. Your information about which social networks you are logged into while browsing.[30]
8. Your installed fonts and which language you are using on your operating system.
9. Your image data.
10. In addition, other technical data, including your screen size, touchscreen support, user agent, status of the Do Not Track (DNT) header, and more.

5.1.3   Web Browser Extensions and Add-Ons

Google Chrome and Mozilla Firefox supply straightforward ways to combat this including the use of add-on "extensions" which serve you by building layers of security into those browsers.

- Visit the articles at the URLs below for advice about these types of extensions then view the sample user extension setups for Chrome and Firefox to get a feel for how you can control collection on your 10 data types.
- NOTE: The Guide suggests adding the extension found below as Protect My Choices **first** on all your browsers then adding Ghostery **second** followed by others.
- This order will first opt your browsers out of interest-based advertising (aka online behavioral advertising) then, second, protect them by blocking tracking ads altogether.
- Be sure to test your browser after setup of add-on extensions to detect any continued unwanted collection or transmission of your data.
- This can be done via open-source tools like Webkay (What Every Browser Knows About You) and Panopticlick; the URLs for these websites are available in Section 3.1.2.

5.2   ONLINE BEHAVIORAL ADVERTISING

Also called "Interest-based advertising", online behavioral advertising targets users with ads based on third-party predictions of their interests and preferences.[31] These predictions are based upon data collected from their devices' web viewing behavior over time and across non-affiliated websites.

- You can control some of this collection via your web browser's privacy controls, by choosing to Opt Out from the online behavioral advertising services run by the Network Advertising Initiative (NAI) and Digital Advertising Alliance (DAA), and by resetting your mobile advertising identifier (see section 3.6.1).[32]
- Further information about online behavioral advertising is available at the Association of National Advertisers at URL *https://www.ana.net/about*.
- You can also enable your browser to run a privacy tool like Ghostery, which blocks advertising attempts to gain access to your information. Ghostery can be read about at the URL https://www.ghostery.com/[33]

5.2.1 Browser Privacy Controls

Platform	Browser	Privacy Advice
Desktop	Chrome	http://support.google.com/chrome/bin/answer.py?hl=en&answer=95647
	Firefox	http://support.mozilla.org/en-US/kb/Enabling and disabling cookies#w_how-do-i-change-cookie-settings
	Internet Explorer	http://windows.microsoft.com/en-US/internet-explorer/delete-manage-cookies#ie=ie-11
	Safari	https://support.apple.com/guide/safari/manage-cookies-and-website-data-sfri11471/mac
	Opera	http://blogs.opera.com/news/2015/08/how-to-manage-cookies-in-opera/
Mobile	Chrome	https://support.google.com/chrome/answer/2392709?hl=en
	Firefox	https://support.mozilla.org/en-US/kb/clear-your-browsing-history-and-other-personal-dat
	Internet Explorer	http://www.windowsphone.com/en-us/how-to/wp7/web/changing-privacy-and-other-browser-settings
	Safari	https://support.apple.com/en-us/HT201265
	Opera	http://blogs.opera.com/news/2015/08/how-to-manage-cookies-in-opera/
	Silk	http://www.amazon.com/gp/help/customer/display.html?nodeId=201730580
	Android Browser	**Click top right corner with three dots, Settings, Privacy**

5.2.2 Online Behavioral Advertising Services

Service	Opt Out
Network Advertising Initiative (NAI)	http://optout.networkadvertising.org/?c=1
Digital Advertising Alliance (DAA) WebChoices Tool	http://www.aboutads.info/choices/
AppChoices (Mobile Apps)	http://www.aboutads.info/appchoices

6 PRIMARY DATA BROKERS

Data brokers collect and sell data about consumers.[34] [35] They do not have a direct relationship with anyone they collect about, but they do sell data to other parties, like companies or individual marketers, for their commercial purposes.[36,37,38,39,40, 41] Primary data brokers sell data to other data brokers.[42]

Primary Data Broker	Opt Out Method
Acxiom	https://isapps.acxiom.com/optout/optout.aspx
CoreLogic	https://www.corelogic.com/privacy-policy/ It is recommended you contact them via the email **privacy@corelogic.com** and you can provide them with documentation to opt out available at URL https://www.corelogic.com/downloadable-docs/teletrack-out-opt-form.pdf
Oracle Data Cloud	https://datacloudoptout.oracle.com/optout/
Epsilon	1. Email optout@epsilon.com; or, 2. Call 1-888-780-3869; or, 3. Send mail to Epsilon, P.O. Box 1478, Broomfield, CO 80036
AddThis	https://www.addthis.com/privacy/email-opt-out
Data and Marketing Choice	https://dmachoice.thedma.org/register.php (Please note that DMA is now charging a $2 fee to register online. If you do not wish to pay $2, you can use the following URL https://dmachoice.thedma.org/prefill_mailin_registration.php to fill out a form and mail your request into DMA.)
Direct Mail	http://www.directmail.com/mail_preference/
E-Bureau	http://www.ebureau.com/privacy-center/opt-out for Opting Out will now route you to TransUnion's Opt Out link. It should be noted that older Opt Out guidance lists Opting Out of E-Bureau so simply Opt Out through TransUnion.
Experian	https://www.experian.com/privacy/opting_out.html
Opt Out Prescreen	https://www.optoutprescreen.com/selection
TowerData	https://instantdata.towerdata.com/optout/
TransUnion Consumer	https://www.transunion.com/customer-support/marketing-offers-opt-out

6.1 REAL ESTATE ONLINE LISTINGS

You should consider removing pictures of your home from real estate services' online listings. These often display both exterior and interior images of your residence.

- Further privacy can be achieved by suppressing curbside images of your home from showing in Google Street View and Bing Curbside. More advice can be at this URL *https://www.thebalance.com/remove-old-home-photos-from-real-estate-websites-4102195*

6.1.1 Real Estate Online Service Privacy Links

Service	Privacy Settings
Zillow	https://zillow.zendesk.com/hc/en-us/articles/218578357-Owner-Dashboard https://zillow.zendesk.com/hc/en-us/requests/new
Trulia	https://support.trulia.com/hc/en-us/requests/new
Realtor	Sign up, control of listing
Redfin	https://support.redfin.com/hc/en-us/articles/360013247432-Removing-Photos-on-a-Sold-Home
Movoto	Contact customercare@movoto.com
Homesnap	Contact support@homesnap.com

6.1.2 How to Remove Curbside Pictures of Your Home

Service	Privacy Settings
Google Street View	https://www.wikihow.com/Opt-Out-of-Google-Street-View https://support.google.com/websearch/answer/4628134?hl=en
Bing Streetside	https://www.bing.com/maps/privacyreport/streetsideprivacyreport?bubbleid=198628406

7 SOCIAL SECURITY NUMBER

The following information from the Social Security Administration (SSA) explains how the nine-digit SSAN (aka SSN) is composed of three parts. More available at URL
https://www.ssa.gov/history/ssn/geocard.html

- The first set of three digits is called the **Area Number**.
- The second set of two digits is called the **Group Number.**
- The final set of four digits is the **Serial Number.**

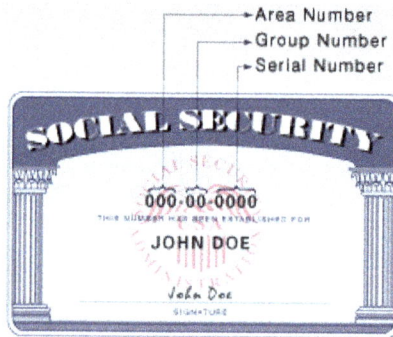

7.1.1 Area Number

The Area Number is assigned by the geographical region. Prior to 1972, cards were issued in local Social Security offices around the country and the Area Number stood for the State in which the card was issued. This did not necessarily have to be the State where the applicant lived, since a person could apply for their card in any Social Security office.

- Since 1972, when SSA began assigning SSNs and issuing cards centrally from Baltimore, the area number assigned has been based on the ZIP code in the mailing address provided on the application for the original Social Security card.
- The applicant's mailing address does not have to be the same as their place of residence. Thus, the Area Number does not necessarily stand for the State of residence of the applicant, prior either to 1972 or since.
- Numbers were assigned beginning in the northeast and moving westward.
- Therefore, people on the east coast have the lowest numbers and those on the west coast have the highest numbers.
- In 2007, the SSA gave public notice that *it intended to abandon its previous method for choosing Social Security numbers and instead to go to a random process for assignment*. The SSA followed through with that change in June 2011.

7.1.2 Group Number

Within each area, the group number (middle two (2) digits) range from 01 to 99 but are not assigned in consecutive order.

- For administrative reasons, group numbers issued first consist of the ODD numbers from 01 through 09 and then EVEN numbers from 10 through 98, within each area number distributed to a State.
- After all numbers in group 98 of a particular area have been issued, the EVEN Groups 02 through 08 are used, followed by ODD Groups 11 through 99.

7.1.3 Serial Number

Within each group, the serial numbers (last four (4) digits) run consecutively from 0001 through 9999. When the government introduced the Social Security program with its numbers in 1936, it was never meant to be so widely used to find and track individuals.

- Today, this number is used for everything from its original purpose – to track your lifetime earnings and calculate your Social Security benefits – to opening a *checking account* or fill out a new-patient form at the doctor's office.
- In the United States, many businesses will ask for your Social Security number simply because it is a convenient way for them to find customers.
- Unfortunately, threat actors can use your Social Security number to commit identity theft, so you should always guard your Social Security number carefully and only give it out when necessary.

7.2 PROTECTING YOUR SSN

Now that you understand what makes up an SSN, here are some simple ways you protect your SSN:

7.2.1 Offer an Alternative Form of Identification

If a business or organization asks for your Social Security number, offer your driver's license number instead.

- Other alternative forms of ID include a passport, proof of current and earlier address (bills) or even a student ID from a college or university.

7.2.2 Ask Why and How the SSN Will Be Handled

If the business insists, ask questions. You have a right to know why it is necessary to supply your SSN and how it will be handled. Here are some questions:

- Why is having my SSN necessary?
- With whom will you share my SSN with if I provide it?
- How will my SSN be stored? Will it be encrypted?
- Do you have a privacy policy, and may I see it?
- Will you cover my *liability* or losses if my SSN is stolen or compromised?
  - Unfortunately, if you are asked to supply your SSN by a business or institution that does not need it and you say no, it can refuse to supply services to you or put conditions on

the service—such as a deposit or added fees. However, the question to always ask is "do I want to do business with a business that does not care about my privacy concerns?"

7.2.3   Leave Your Card at Home
Do not carry your card around with you in your wallet or purse.

- Do not enter it into your phone, laptop, or other device. It is unlikely you will need your card and when you do need it, it does not come as a surprise.

7.2.4   Shred Mail and Documents with Personal Details
Discarded mail and documents are easy places for identity thieves to search. Do not just throw out papers that hold personal details such as your SSN.

- Get a shredder at a discount or office supply store and use it on a regular basis.
- Do not leave mail in an outside mailbox for prolonged periods. Stealing mail is another way a thief can make off with your identity.

7.2.5   Do Not Use Your SSN as a Password
Do not use the whole number—or part of it—as a password for anything! The password file can be stolen and decrypted, or someone can just watch you type it in from over your shoulder.

- Also, if you need to require it for legitimate purposes in a public place, be careful who may be able to eavesdrop on your conversation.

7.2.6   Do Not Send Your SSN via Electronic Device
Never type your SSN into an email or instant message and send it. Most email messages can be intercepted and read in transmission.

- Also, do not leave a voice mail that includes your SSN. If you need to contact someone and give them your number, it is always best to do so in person.
- If you need to do so on the phone, ensure you are speaking to the right person, so you are not swindled.

7.2.7   Do Not Give Your SSN Out
You should never supply your SSN to someone you do not know who calls you on the phone and requests it. This same warning applies to unsolicited emails and any forms you fill out on the internet.

- In general, do not give your SSN to anyone unless you are certain they have a reason and a right to have it.

7.2.8   Monitor Bank and Credit Card Accounts
Keep close tabs on your bank and credit card balances.

- This is one way to make sure your SSN and *identity have not been compromised*.

- Many banks let you sign up for account alerts. They will send you text alerts or call you if transactions exceed a certain amount or if someone tries to use your SSN to access your account.
- You can also check your *credit score* on a regular basis at *AnnualCreditReport.com*. You can do this once a year free.
- If the *Social Security Administration* is still sending you an annual statement detailing your earnings, and it looks abnormal, someone might be using your SSN for employment purposes. You can register to get statements at the Social Security Administration's *website*.

7.2.9   Use an Identity Protection Service

You can register with (and pay for) an identity protection service such as *LifeLock*, *IdentityForce*, or *Identity Guard*.

- Such services supply identity insurance—for a fee, that typically starts around $10 per month.
- Banks and credit unions also have packages they sell to customers, as do major *credit rating* agencies such as Experian and TransUnion.

7.2.10   Do Not Forget to Protect Your Child's SSN

- While you are protecting your own Social Security number, make sure you are equally watchful about your children's numbers.

7.2.11   Block Access to Your SSN

- Electronic and phone access to SSN information can be blocked by going to the *Block Electronic Access page* on the Social Security Administration's website. Once there, you will verify your identification and confirm your intention to block your Social Security number.
- Blocking your number will prevent access by anyone, including you. If conditions change or you need access to your information, the block can be lifted either permanently or temporarily by contacting the *Social Security Administration*.
- If you would like more information about the benefits of blocking your SSN, read the article at URL *https://www.sapling.com/6926296/block-social-security-number*

Service	URL
SSA Block Electronic Access Page	https://secure.ssa.gov/acu/IPS_INTR/blockaccess
SSA Contact Page	http://www.socialsecurity.gov/agency/contact/
Office of the Inspector General: SSA Scam Reporting Form	https://secure.ssa.gov/ipff/home

7.2.12   E-Verify

E-Verify, authorized by the Illegal Immigration Reform and Immigrant Responsibility Act of 1996 (IIRIRA), is a web-based system through which employers electronically confirm the employment eligibility of their employees.

- E-Verify is administered by SSA and U.S. Citizenship and Immigration Services (USCIS). USCIS facilitates compliance with U.S. immigration law by supplying E-Verify program support, user support, training, and outreach, and developing innovative technological solutions in employment eligibility verification.[43]

7.2.13 E-Verify Self Lock

Self-Lock is the unique feature that lets you protect your identity in E-Verify and Self Check by placing a "lock" on your Social Security number (SSN).

- This helps prevent anyone else from using your SSN to try to get a job with an E-Verify employer.
- If an employer enters your locked SSN in E-Verify to confirm employment authorization, it will result in an E-Verify mismatch, called a **Tentative Nonconfirmation** (TNC).[44]

7.2.13.1 Using E-Verify Self Lock

To access Self Lock, you must be logged in to your myE-Verify account. To lock your SSN, you must enter your SSN and date of birth. myE-Verify does not store your SSN when you create your account, so you must supply your SSN to "lock" it.

- In addition, you must select and answer three challenge questions. Select questions you can easily answer, because you will need to answer them again to verify your identity if you receive an *E-Verify Tentative Nonconfirmation* due to Self-Lock.

8 PEOPLE SEARCH SITES

People search sites enable the public to search names and other personally identifiable information.[45, 46, 47, 48] Returns from these searches include property addresses, points of contact, family members, aliases, and more associated with the searched information with varying degrees of accuracy.

People Search Site	Opt Out Method
Addresses	https://www.intelius.com/opt-out/submit/
Archives	http://www.archives.com/?_act=Optout
BeenVerified	https://www.beenverified.com/f/optout/search
Cubib	https://cubib.com/optout.php
FamilyTreeNow	https://www.familytreenow.com/optout
FastPeopleSearch	https://www.fastpeoplesearch.com/removal
Instant Checkmate	https://www.instantcheckmate.com/opt-out/
Intelius	https://www.intelius.com/optout
Lexis Nexis	https://www.lexisnexis.com/en-us/privacy/for-consumers/opt-out-of-lexisnexis.page?
Peek You	https://www.peekyou.com/about/contact/optout/
People Finders	https://www.peoplefinders.com/opt-out
People Smart	https://www.peoplesmart.com/optout-go
People Wiz	https://www.peoplewhiz.com/remove-my-info
Pipl	https://pipl.com/help/remove/
Radaris	https://radaris.com/ng/page/removal-officer
Social Catfish	https://socialcatfish.com/opt-out/
Spokeo	https://www.spokeo.com/optout
SpyFly	https://www.spyfly.com/help-center/remove-info
ThatsThem	https://thatsthem.com/optout
TruePeopleSearch	https://www.truepeoplesearch.com/removal
USA People Search	https://www.usa-people-search.com/manage/
White Pages	https://www.whitepages.com/data-policy
USPhoneBook	http://www.usphonebook.com/opt-out

9 MOBILE PHONES AND MOBILE BROWSING

For most of us, our mobile phone is the single most valuable tool we carry, but malicious actors can also use it against us.[49] It is important to know what your phone holds[50] and how it can make you vulnerable to attacks.[51]

- Mobiles phones have a variety of sensors and software, which generate data useful for finding and tracking you.[52,53,54,55, 56]
- Check your location settings and advertisement settings via advice below. Be aware smartphone apps could also leak your personal data to include your location.[57,58,59,60, 61]
- Privacy advice for safely downloading smartphone apps can be read at the URL *https://www.cnet.com/news/7-privacy-tips-for-safely-using-smartphone-apps/* and below for Apple and Android technology settings.

9.1.1 Mobile Phones

Technology	Privacy Advice
Location Settings	https://www.digitaltrends.com/mobile/android-privacy-guide/ *Pixel only:* https://android.gadgethacks.com/how-to/20-privacy-security-settings-you-need-check-v google-pixel-0193251/
Limit App Store Interest-based Ads	https://support.google.com/googleplay/android-developer/answer/6048248?hl=en#zippy=%2Chov out-of-personalized-ads
Limit Ad Tracking	https://support.google.com/accounts/answer/2662856?co=GENIE.Platform%3DDesktop&oco=1#e
Location Settings	https://support.apple.com/en-us/HT207092
Limit Ad Tracking	https://support.apple.com/en-us/HT202074
Limit App Store Interest-based Ads	https://support.apple.com/en-us/HT202074
Reset Mobile Advertising Identifier	https://www.adcolony.com/privacy-policy/finding-advertising-id/

9.2 iPhone Privacy Settings

Apple in June 2021 introduced the latest version of its iOS operating system, iOS 15, which was released in September 2021. Apple's iOS 15 is the latest version of the mobile operating system and features several new privacy features that were not previously available with older operating systems. The newest privacy features are as follows:

9.2.1 Custom Alphanumeric Code

With the rollout of iOS 15, you can now generate a strong passcode using Custom Alphanumeric Code if you suspect someone knows your passcode.[62] To do so, complete the following steps:
- Go to **Settings > Face ID & Passcode** (or Touch ID & Passcode).
- Turn on **Face ID/Touch ID**.
- Turn on screen **Auto-Lock**.
  Go to **Settings > Display & Brightness** and tap **Auto-Lock** and set to 30 seconds or 1 minute.
- Make sure iOS is up to date.
  Go to **Settings > General > Software Update** and make sure **Automatic Update** is enabled.
- Keep all your apps updated.
  Go to **Settings > App Store** and make sure **App Updates** are enabled.

9.2.2 Record App Activity

A new feature in iOS 15 is the ability to log what apps are up to on your iPhone. The feature is called *Record App Activity*, and this allows you to get a lot of when an app does one of the following[63]:

- The user's photo library
- A camera
- The microphone
- The user's contacts
- The user's media library
- Location data
- Screen sharing
- To enable this feature, go to **Settings > Privacy** and then scroll down to find **Record App Activity**.

9.2.3 Built-in Authenticator

With the rollout of iOS 15, users have the option to use a built-in authenticator rather than simply choosing to use a third-party two-factor authenticator app.[64] If you choose to use this feature, simply follow the steps below:
- Got to **Settings > Passwords,** and then for each password entry, you can tap on it to get access to a choice called **Set Up Verification Codes...** which allows you to enter the information needed either using a setup key or QR code.
- Using a two-factor authenticator is far more secure than relying on SMS messages, so you should use this feature either using Apple's authenticator or another app to get the highest security.

9.2.4 Hide Your IP Address from Trackers

Safari can now cloak your IP address from trackers on websites, making it impossible for your browsing to be logged.[65]

- Go to **Settings > Safari** and set **Hide IP Address** to **From Trackers**.

9.2.5 Secure your browsing

If you have an iCloud+ subscription, Apple has just given you a great reason to use the Safari browser -- iCloud Private Relay. This is like a VPN in that it sends your web traffic through other servers to keep your location secret.[66]

- To enable iCloud Private Relay, you will need an iCloud+ subscription.
- Then go to **Settings,** and at the top, tap your name and then go to **iCloud** and enable **Private Relay**.

9.2.6 Stopping Email Trackers

Protect Mail Activity is a feature built into the Mail app that prevents people from knowing if emails have been opened.[67]

- To enable this feature, go to **Settings > Mail**, tap on **Privacy Protection** and enable **Protect Mail Activity**.
- If iCloud Private Relay is a good reason to switch to Safari, then this feature is a good reason to switch to Mail.

9.2.7 Privacy-focused Apple Calendar Settings

While there is no known open-source reporting about Apple calendars being used by threat actors to target users through the creation of messages used in phishing schemes or social engineering attacks, the following URLs will help you ensure your Apple Calendars are configured properly.

Browser Privacy Control	URL
Apple Calendar (Share Calendars)	https://support.apple.com/kb/PH2690?locale=en_US
Apple Calendar (Stop Sharing Calendars)	https://support.apple.com/guide/icloud/stop-sharing-a-calendar-mm6b1a8f9f/icloud

9.2.8 iPhone App Store Personalized Recommendations

Click on the **Account Settings** button, which will prompt you for your passcode or a biometric identifier. Once in, look for the setting entitled **Personalized Recommendations**.

- If the switch is green, the settings is *enabled,* and you iPhone will send you Personalized Recommendations. Ensure the switch is not green to *disable* this feature.
- Apple describes Personalized Recommendations as ""*when you download from a Store, or install an app on your Apple Watch, identifiers such as Apple logs your device's hardware ID and IP address along with your Apple ID. Apple further describes that they find ways use information about your browsing, purchases, searches, and downloads. These records are stored with IP*

address, a random unique identifier (where that arises), and Apple ID when you are signed into a Store "at the URL https://support.apple.com/en-us/HT208477

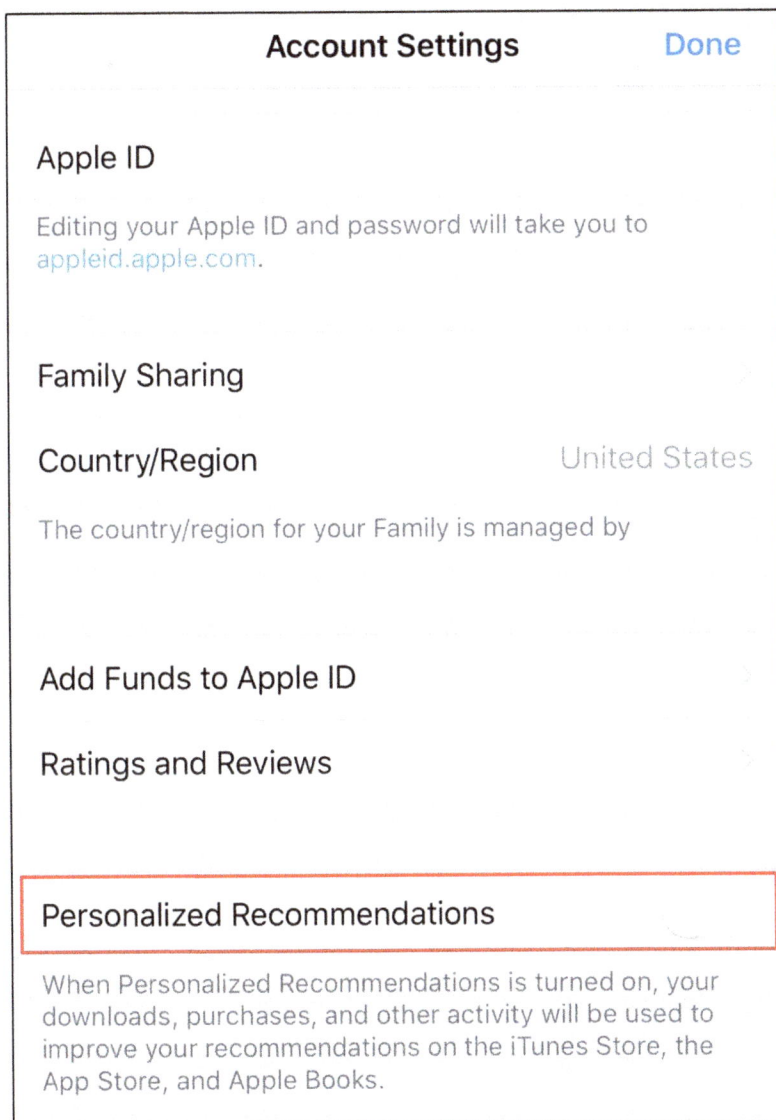

Account Settings   Done

Apple ID

Editing your Apple ID and password will take you to appleid.apple.com.

Family Sharing

Country/Region   United States

The country/region for your Family is managed by

Add Funds to Apple ID

Ratings and Reviews

Personalized Recommendations

When Personalized Recommendations is turned on, your downloads, purchases, and other activity will be used to improve your recommendations on the iTunes Store, the App Store, and Apple Books.

9.2.9   Country/Region Settings

It is important to note that US users should ensure the Country/Region is set to the United States and not set to a different country.

- A misconfiguration of these setting risks having all your account's data transferred to another country beyond the protections afforded by the US Constitution AND may directly expose it to threats from any government who's Intelligence or Law Enforcement services may or may not have means to decrypt what is stored in their country.
- Additional information on tips on how to ensure your safety when traveling to high-risk areas can be found at the URL *https://travel.state.gov/content/travel/en/international-travel/before-you-go/travelers-with-special-considerations/high-risk-travelers.html*.

9.3   IPHONE ADS AND LOCATION SETTINGS

This section guides you how to control your iPhone's **Analytics** and **Advertising**, **Location Services**, ability to deliver **Location-based Apple Ads**, track your **Significant Locations**, and ability to deliver **Personalized Recommendations** through your location. This *URL* will inform you how your iPhone shares analytics, diagnostics, and usage information with Apple.

- With the rollout of Apple's new *iOS 15*, the following tips are still applicable though users have a greater ability to manipulate privacy settings within iOS 15. This *URL* will inform you on some key features within iOS 15 that will better enhance your iPhone Analytics.

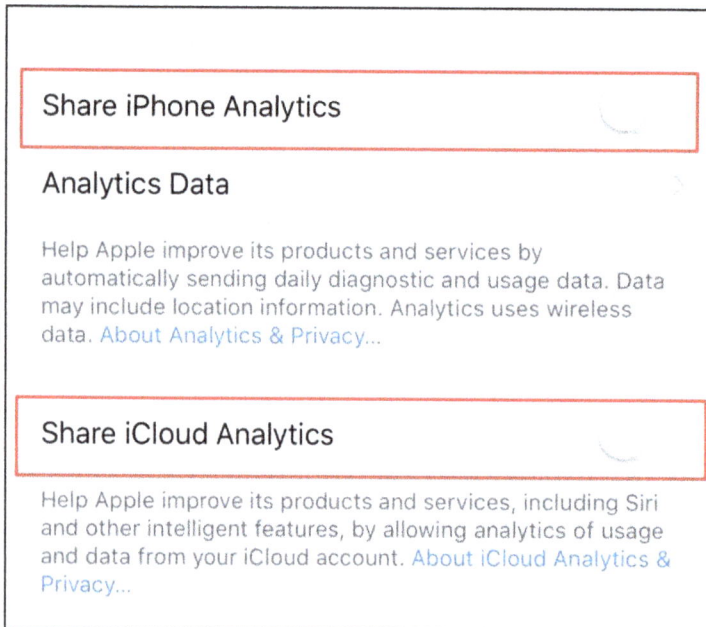

Share iPhone Analytics

Analytics Data

Help Apple improve its products and services by automatically sending daily diagnostic and usage data. Data may include location information. Analytics uses wireless data. About Analytics & Privacy...

Share iCloud Analytics

Help Apple improve its products and services, including Siri and other intelligent features, by allowing analytics of usage and data from your iCloud account. About iCloud Analytics & Privacy...

9.3.1   iPhone Advertising

Click on the "**Reset Advertising Identifier**" section periodically to ensure you are controlling what Apple describes as "Segments" of your personal information and data.[68] If you would like to know more about the information used by Apple to deliver relevant Apple ads to you in Apple News and the Apple App Store, click the "**View Ad Information**" section to view your personalized data.

- You can read more about "segments" at the URL *https://support.apple.com/en-us/HT205223*

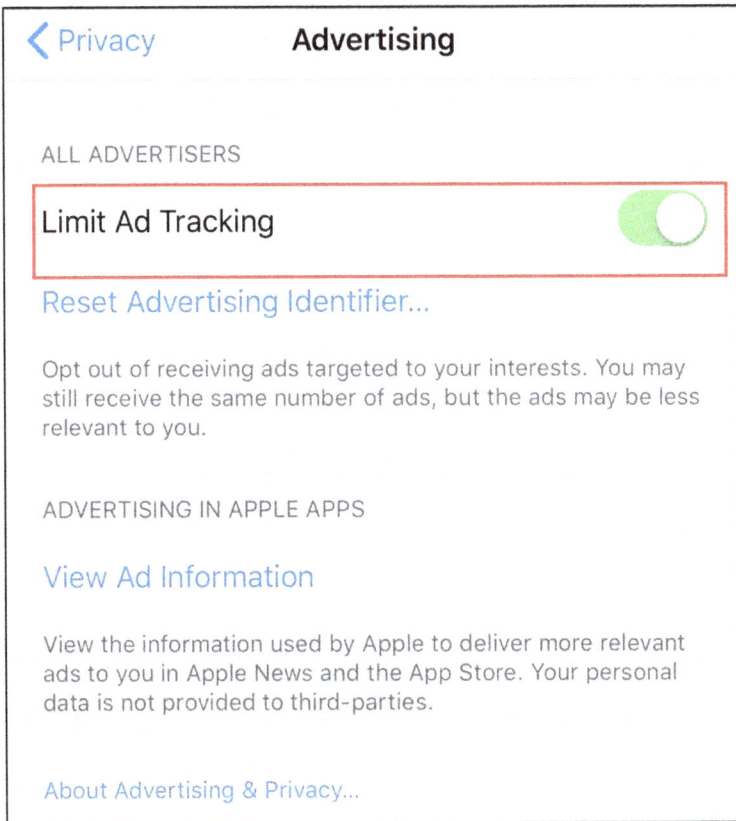

ALL ADVERTISERS

Limit Ad Tracking

Reset Advertising Identifier...

Opt out of receiving ads targeted to your interests. You may still receive the same number of ads, but the ads may be less relevant to you.

ADVERTISING IN APPLE APPS

View Ad Information

View the information used by Apple to deliver more relevant ads to you in Apple News and the App Store. Your personal data is not provided to third-parties.

About Advertising & Privacy...

9.3.2 iPhone Location Services

Open Settings and tap Privacy. You will now see the Location Services as shown in the graphic. According to Apple, Location Services uses GPS and Bluetooth (where available), along with crowd-sourced Wi-Fi hotspots and cellular towers to find the approximate location of your device.[69]

- The website also describes Apps won't use your location until they ask for your permission and you allow permission." Review this for yourself at the URL *https://support.apple.com/en-us/HT207092* [70]
- Click on **Location Services** and you will see all the Apps your phone has installed and what type of access you have given each App about using your iPhone's location. You have three options available**: "Always", "While Using The App"** and **"Never"**.
- What setting you use depends on your preferences so after you evaluate your App location settings, scroll to the bottom of the page, and look for System Services, as shown in the graphic.

9.3.3   iPhone Location-based Apple Ads

System Services	◹ ›

◹ A hollow arrow indicates that an item may receive your location under certain conditions.

◹ A purple arrow indicates that an item has recently used your location.

◹ A gray arrow indicates that an item has used your location in the last 24 hours.

Then,

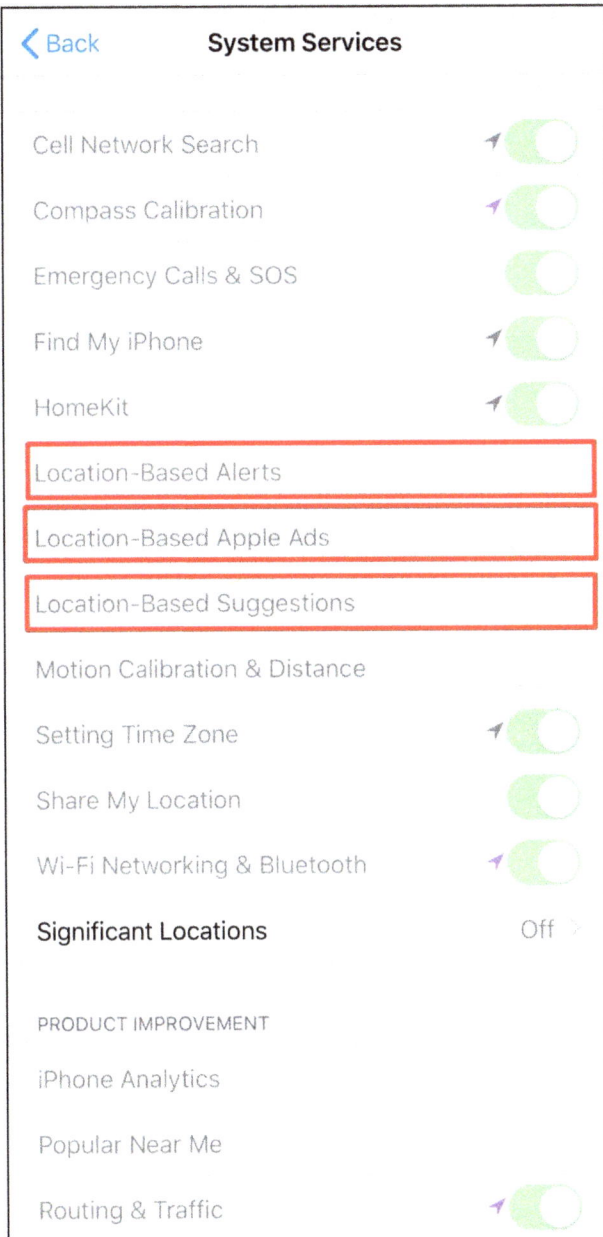

‹ Back	**System Services**

Cell Network Search

Compass Calibration

Emergency Calls & SOS

Find My iPhone

HomeKit

Location-Based Alerts

Location-Based Apple Ads

Location-Based Suggestions

Motion Calibration & Distance

Setting Time Zone

Share My Location

Wi-Fi Networking & Bluetooth

Significant Locations   Off ›

PRODUCT IMPROVEMENT

iPhone Analytics

Popular Near Me

Routing & Traffic

9.3.4 iPhone Significant Locations
The **Significant Locations** setting allows your iPhone to keep track of places you have recently been as well as how often and when you visited them.[71]

- Apple explains these data are "*encrypted and stored only on your device and will not be shared without your consent. It is used to provide you with personalized services, such as predictive traffic routing, and to build better Photos Memories*" at the URL *https://support.apple.com/en-us/HT207056*

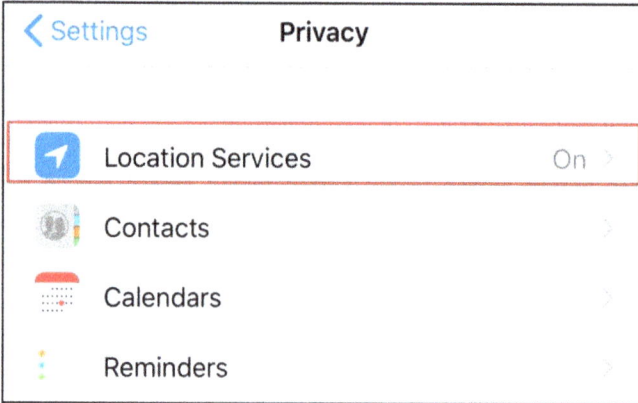

9.3.5   Find My

Within iOS 15, the Find My app introduces new abilities to help locate a lost device that has been turned off or erased using the **Find My network**.[72] Any trusted connections to a user can share their location with which will continuously live-stream their location to provide a sense of direction and speed.[73] There are also new Separation Alerts to notify a user if they leave an AirTag, Apple device, or Find My accessory network behind in an unfamiliar location.[74]

9.4   ANDROID PRIVACY SETTINGS

Your Android phone includes records of everywhere you go alongside most, if not all, of your digital communication and Internet search history.[75] The following section is designed to help users to understand and adjust privacy settings and reduce their Digital Exhaust.[76]

9.4.1   Connected devices

- On the *Settings* screen, tap **Connected devices**.
- If there are any connections you are not using right now, such as Bluetooth, tap them and toggle them **off**. Only enable connections when you truly need them. This limits the ways your device could be compromised and limits how your location can be tracked.[77]

9.4.2   Apps & Notifications

- In the top left, tap the **back arrow** until you are back to the *Settings* screen. Then, tap **Apps & notifications**.
- Tap **See all # apps**. Go through the *App info* list and for any that, you do not truly need, tap the app, and then tap **Uninstall**. Many pre-installed apps cannot be uninstalled, so you will not see

an *Uninstall* button. For those, you can tap **Disable** to turn the app off and hide it from your device.[78]

- In the top left, tap the **back arrow**. Then, tap the **Permission manager**. Tap each permission (Body sensors, Calendar, etc.) to see the apps with that permission. If any app should not have the permission, tap it, and then tap **Deny**.
- In the top left, tap the **back arrow**. Then, tap **Advanced**, then **Emergency alerts**. Toggle **on** any emergency alerts you want to receive.

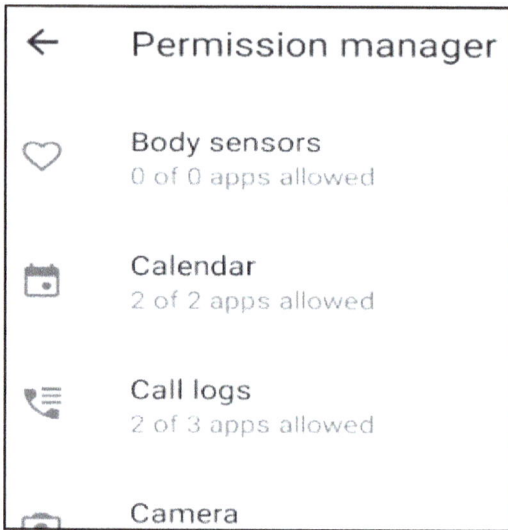

Permission manager

← Permission manager

♡ Body sensors
0 of 0 apps allowed

📅 Calendar
2 of 2 apps allowed

📞 Call logs
2 of 3 apps allowed

📷 Camera

← Contacts permission 🔍

Chrome

CONTACTS ACCESS FOR THIS APP

○ Allow

◉ Deny

See all Chrome permissions

9.4.3   Display

- In the top left, tap the **back arrow** until you are back to the *Settings* screen. Then, tap **Display**.
- Tap **Screen timeout**. Choose a brief time (I recommend *1 minute* or less). When you add a screen lock later, this will cause the screen to lock after a brief period, preventing others from using your device.

- Go back to the *Display* screen, then tap **Advanced**, and then **Lock screen display**, then **Lock screen**. I recommend choosing **Don't show notifications at all**, because notifications can reveal sensitive data (messages, calendar reminders, etc.).
- Tap **Lock screen message**. Here you can set a message that shows on the lock screen. If a Good Samaritan finds your device, this will tell them how to contact you. However, do not give away too much personal info, because a nefarious person could use it against you. Do not put your home address. I recommend putting a phone number and/or email address.

9.4.4   Android Privacy
- In the top left, tap the **back arrow** until you are back to the *Settings* screen. Then, tap **Privacy**.[79]
- Tap **Autofill service from Google**, if you want your device to automatically fill in personal info, addresses, and passwords for you. If you previously enabled this and now want to disable it, I will tell you how in the *System* section.[80]
- Go back to the **Privacy** screen, then tap **Advanced**, then **Activity controls**. I recommend that you toggle **off** as many as possible; to reduce the amount of data Google collects about you. I cover these controls in the *Google Account Security & Privacy Guide*.
- Go back to the **Privacy** screen, and then tap **Ads**. Toggle **on** *Opt out of Ads Personalization* to reduce the amount of data Google collects about you.
- Go back to the **Privacy** screen, and then tap **Usage & diagnostics**. I like to share data that helps make software and services better if my data is anonymized. If you prefer, you can toggle **Off**.

9.4.4.1   Location
- In the top left, tap the **back arrow** until you are back to the *Settings* screen. Then, tap **Location**.[81]
- If you do not want to use the location at all, you can toggle **off** *Use location*. Note that location must be on for *Find My Device* to work (which lets you remotely find, lock, and wipe/erase your device).[82]
- Tap **Wi-Fi and Bluetooth scanning**. I recommend toggling these **off** unless you truly need exact locating. If you toggle these on your device can use Wi-Fi and Bluetooth signals for location, even when you have turned off Wi-Fi and Bluetooth.[83]

9.4.4.2   Android Security
- In the top left, tap the **back arrow** until you are back to the *Settings* screen. Then, tap **Security**.[84]
- Tap **Google Play Protect**, then the **gear icon** in the top right. Toggle **on** *Scan apps with Play Protect* and *Improve harmful app detection*.[85]

Play Protect settings

General

Scan apps with Play Protect
Play Protect can scan this device and warn you about harmful apps

Improve harmful app detection
Send unknown apps to Google for better detection

- Go back to the **Security** screen, and then tap **Find My Device**. It is recommended toggling this **on**. It allows you to remotely find, lock, and wipe/erase your device if it becomes broken, lost, or stolen.
- Go back to the **Security** screen, and then tap **Security update**, if you see it. If it shows an available update, install it.
- Go back to the **Security** screen, and then tap **Screen lock**. Setting a password is best, but because it is annoying to type a password on a mobile device, consider setting a pattern or PIN. Ensure the pattern is complex, and the PIN is at least 6 digits (the longer, the better).
- Go back to the **Security** screen, and then tap **Fingerprint**. You can choose to use your fingerprint along with another screen lock method.
- Go back to the **Security** screen, and then tap **Advanced**, then **Encryption & credentials**. If you do not see **Encrypted** under **Encrypt phone**, then tap it to enable encryption. Encrypting your device is one of the best things you can do to secure it, because it means that if someone steals your device, they will not be able to see or copy your data off the device.

9.4.4.3   Text (SMS) Message Security
- *Text (SMS) messages are not secure*. If you are communicating about anything sensitive or confidential, you consider a secure, private messaging app.

9.4.4.4   Accounts
- In the top left, tap the **back arrow** until you are back to the *Settings* screen. Then, tap **Accounts**.
- Android is meant to be used with a Google account. If you sign into a Google account, you will have many more options. However, you can use an Android device without a Google account. Another choice is to create a separate Google account that you use just for Android, and do not use it for anything else.

- You can toggle **Automatically synchronize data** if you want apps to automatically synchronize with accounts. If you toggle it off, you can still manually synchronize accounts.
- Tap an account, and then tap **Account sync** to customize what is synchronized. Toggle off any items that you do not need to be synchronized to your device.

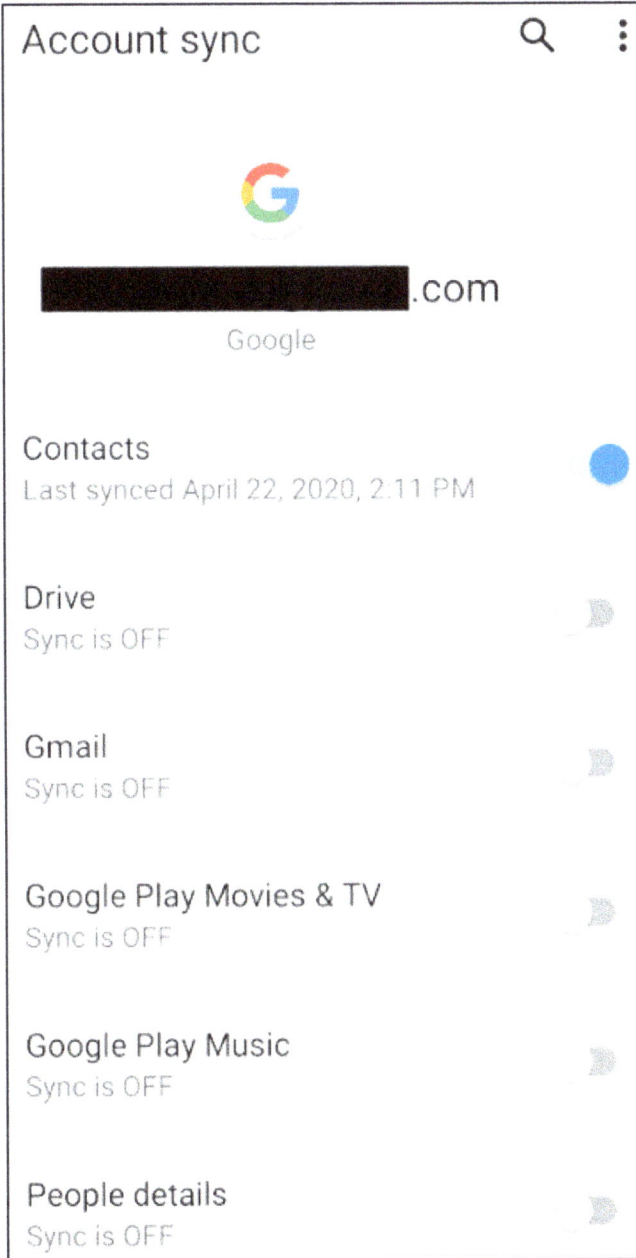

Account sync   Q   ⋮

G

████████████.com
Google

Contacts
Last synced April 22, 2020, 2:11 PM

Drive
Sync is OFF

Gmail
Sync is OFF

Google Play Movies & TV
Sync is OFF

Google Play Music
Sync is OFF

People details
Sync is OFF

9.4.4.5 Android Anti-Malware

- It is always recommended that you use antivirus software to protect your Android device.
- One choice is to manually scan weekly (run an **on-demand scan**), rather than having an anti-malware app run constantly in the background (sometimes called **real-time scanning**).

9.5 GOOGLE ACCOUNT SETTINGS

- In the top left, tap the **back arrow** until you are back to the *Settings* screen. Then, tap **Google**.
- Tap **Account services**, then **Connected apps**. You will see the apps and devices connected to your Google account. If any should be disconnected, tap them, and click **Disconnect**.
- In the top left, tap the **back arrow** until you are back to the **Account services** screen. Then, tap **Search, Assistant & Voice**, and then **Google Assistant**. Google Assistant is, well, Google's digital assistant, the equivalent of Amazon's Alexa and Apple's Siri. To work, Google Assistant sends a lot of data about what you say, type, and do to Google. If you do not want to use it, tap the **Assistant** tab, and scroll down to **Assistant devices**. Tap your device. Then, toggle **off Google Assistant**.
- Anyone who is near your Google speaker or display device can request information from it, and if you have given your device access to your calendars, Gmail or other personal information, people may be able to ask your device about that information, depending on your *personal results settings* and *Voice Match settings.* Google employees and trusted third parties can also access your conversation history in line with Google's *Privacy Policy.*

9.5.1 Google Assistant

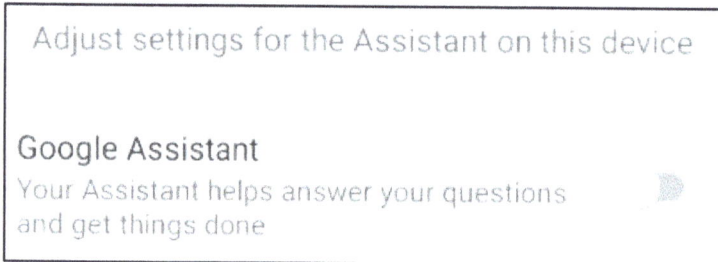

Adjust settings for the Assistant on this device

Google Assistant
Your Assistant helps answer your questions and get things done

- If you want to use Google Assistant, go back to the **Account services > Search, Assistant & Voice** screen and configure the settings in **Google Assistant** and **Voice**.
- If your child will be using this device, you can go back to the *Google* screen and tap **Parental controls** to set up *Google Family Link*. It lets you control content, apps, and screen time.

9.5.2 System

- In the top left, tap the **back arrow** until you are back to the **Settings** screen. Then, tap **System**.
- If you previously enabled **Autofill service from Google** (to automatically fill in personal info, addresses, and passwords) and now want to disable it, tap **Languages & input**, then **Advanced**, then **Autofill service**, then **Autofill service**. Then, select **None**.

- Go back to the **System** screen, and then tap **Backup**. Toggle on **Back up to Google Drive** unless you will be using a different backup service. If you are running Android 9 ("Pie") or later, Google cannot see your backup data.
- If your backups are uploaded in Google, they are encrypted using your Google Account password. For some data, your phone's screen lock PIN, pattern, or password is also used for encryption.

- This decryption key is encrypted using the user's lock screen PIN/pattern/passcode, which is not known by Google. ... By design, this means that no one (including Google) can access a user's backed-up application data without specifically knowing their passcode.

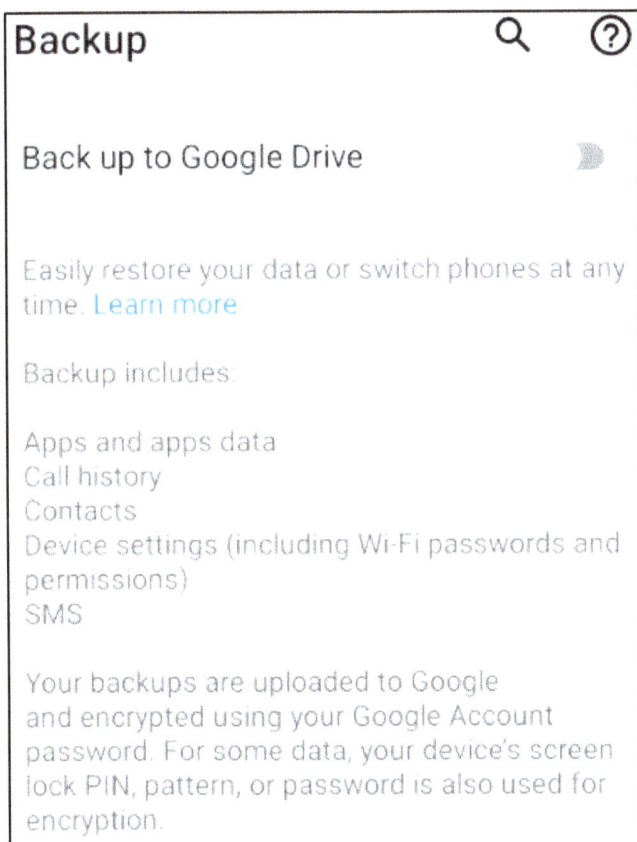

Backup Q ⑦

Back up to Google Drive   ⟫

Easily restore your data or switch phones at any time. Learn more

Backup includes:

Apps and apps data
Call history
Contacts
Device settings (including Wi-Fi passwords and permissions)
SMS

Your backups are uploaded to Google and encrypted using your Google Account password. For some data, your device's screen lock PIN, pattern, or password is also used for encryption.

9.5.3   Updating Apps
- Because app updates often fix security vulnerabilities, you should install them as soon as they are available.[86]
- Open the **Google Play** app, then tap the **menu** (hamburger icon, 3 horizontal lines in the top left), then tap **Settings**, then **Notifications**. Toggle **on Updates.**

- Tap the back arrow in the top right to go back to **Settings**, and then tap **Auto-update apps**. Set it to **Over Wi-Fi only**. If you rarely connect to Wi-Fi, set it to **Over any network**.
- Whenever your device shows that updates are waiting to be installed, install them.

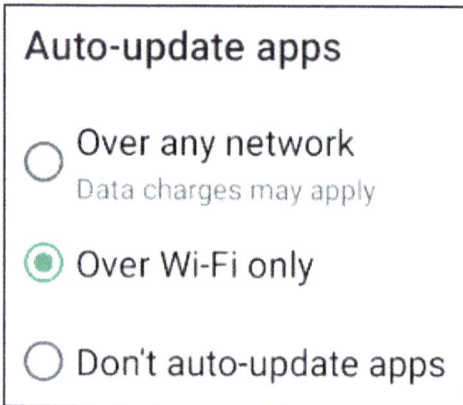

Auto-update apps

○ Over any network
Data charges may apply

◉ Over Wi-Fi only

○ Don't auto-update apps

9.6 MOBILE BROWSING

Online privacy is a major concern in the tech world, and by far some of the biggest privacy issues arise when you browse the internet, even if you use a mobile browser.[87] Having a solid understanding of these privacy settings is critical to reduce your Digital Exhaust, as a user will be exposed to many techniques to track them around the web due to cookies, your IP address, and other device-specific identifiers.[88]

Platform	Technology	Privacy Advice
Browser	Chrome	https://defendingdigital.com/google-chrome-security-privacy-guide/
	Firefox	https://restoreprivacy.com/firefox-privacy/
	Safari	https://defendingdigital.com/apple-safari-security-privacy-guide/
	Brave	https://support.brave.com/hc/en-us/articles/360017989132-How-do-I-change-my-Privacy-Settings-
	Edge	https://privacyinternational.org/guide-step/4333/edge-adjusting-settings-enhance-your-online-privacy
	Opera	https://help.opera.com/en/latest/security-and-privacy/
Search Engine	Google	https://www.cnet.com/google-amp/news/do-you-care-about-online-privacy-then-change-these-browser-settings-immediately/
	DuckDuck Go	https://spreadprivacy.com/how-anonymous-is-duckduckgo/
	Google	https://www.pcworld.com/article/3299042/privacy/google-privacy-checkup-faq.html https://www.pcworld.com/article/3315701/mobile/how-to-delete-google-search-history.html

9.7 MOBILE TWO-FACTOR AUTHENTICATION

- If you do not have two-factor authentication (2FA) enabled yet on your iPhone, consider doing so. This adds another layer of security to your logins by requiring more than just your password.[89]
- These codes often arrive via text or email, though you can get 2FA codes through an app instead. Here is how to enable that feature:

9.7.1 iPhone Two-Factor Authentication
Here is how to enable that feature on an iPhone:

- Go to **Settings** > **[your name]** > **Password & Security** and tap **Turn on Two-Factor Authentication**.
- Tap **Continue**, and then enter the phone number where you want to receive the verification codes.
- Tap **Next** and enter the code.

9.7.2 Android Two-Factor Authentication
Here is how to enable that feature on an Android:

- Open your Google Account and select **Security**.
- Select **2-Step Verification** (under Signing into Google) and then **Get started**.
- Now pick a method for verification: Google prompts, security keys, Google Authenticator or similar apps, or a verification code sent to your phone via text or call.

9.8 GEOLOCATION DATA
Information about where devices are found can serve as a proxy for where individuals are found over time[90], which can be very revealing of an individual's behavior[91], interests[92], or beliefs[93].

- Mobile devices, from smart phones to tablets to fitness trackers, have become intertwined in many people's lives over the last decade, supplying many benefits and becoming almost indispensable.[94]
- However, the benefits and convenience can come at a cost.[95]
- Mobile devices store and share valuable location data by design.[96]
- This data can reveal details about the number of users in a location, user and supply movements, daily routines and can expose otherwise unknown associations between users and locations.[97]

The following graphics give you an overview of how location data is generated, who has access to it, and how is it used.

THE WORLD OF GEOLOCATION DATA

Information about where devices are located can serve as a proxy for where individuals are located over time, which can be very revealing of individual behavior, interests, or beliefs. How is location data generated, who has access to it, and how is it used?

9.9 Stop Contacts From Syncing To Mobile Apps

9.9.1 iPhone Settings

- Go to **Settings,** Screen **Time**, and then **Content & Privacy Restrictions** (as shown in the graphic).

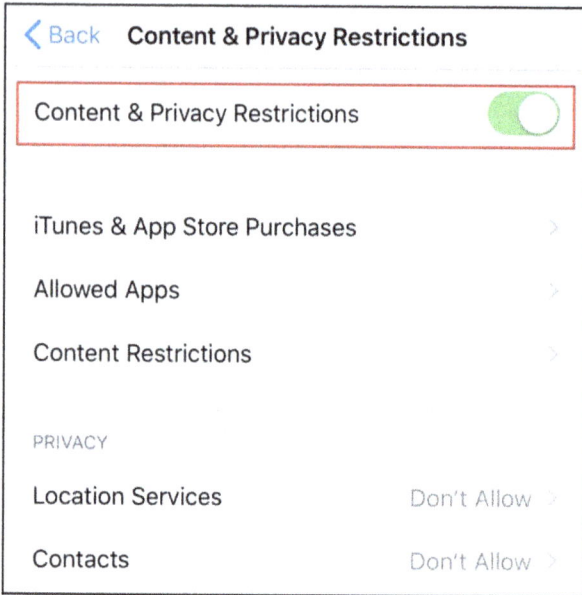

Content & Privacy Restrictions screen showing:
- Content & Privacy Restrictions (toggle, enabled)
- iTunes & App Store Purchases >
- Allowed Apps >
- Content Restrictions >

PRIVACY
- Location Services — Don't Allow >
- Contacts — Don't Allow >

- Then, Enable **Content & Privacy Restrictions**.
- Scroll down the Privacy section and tap on **Contacts.** Tap **Do not Allow Changes** to lock the settings your iPhone's contacts are now locked down from Apps.

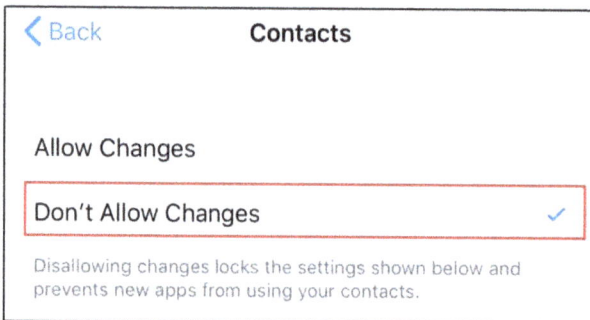

Contacts screen showing:
- Allow Changes
- Don't Allow Changes ✓

Disallowing changes locks the settings shown below and prevents new apps from using your contacts.

9.9.2   Android Settings

Steps may vary depending on which Android Mobile Phone you use, but generally:

1. Open the **Settings** app.
2. Tap the **Apps & notifications** choice.
3. Tap the app you want to examine.
4. Tap **Permissions** to see everything the app can access.
5. To turn off permission, tap on it. You might need to tap a confirmation box here as well.

9.10 Blocking Unwanted Calls

9.10.1 Your Personal Telephone Number

Your personal telephone number is one of your biggest digital exhaust personal vulnerabilities.[98] [99]You can decrease this vulnerability by setting up extra security for the phone. [100]

- If you switch your phone number, often, recycled numbers allow new customers access to old customer information, opening opportunities for a variety of potentially exploitative encounters.[101]
- Create a security code and/or obfuscate the true number by creating a separate forwarding number. Read about this at the URL *https://techcrunch.com/2018/12/25/cybersecurity-101-guide-protect-phone-number/*

9.10.2 iPhone: How to Block a Number

There are multiple methods of how to block a number on iPhone devices. Before following the steps below, make sure your iPhone is updated.

9.10.2.1 Via Your Call History
- Go to your **Phone** icon/app.
- Click on the blue ? symbol next to the restricted call.
- Choose *Block this caller* to block the specific restricted call.

9.10.2.2 Use Do Not Disturb
- Go to Settings → Do Not Disturb.
- Scroll down to Allow Call From and click on that.
- Choose who you want to accept calls from, such as your Favorites or All Contacts.
- On the Do Not Disturb page, make sure your other settings are set the way you want them.
- Turn on the Do Not Disturb button at the top of the page.

9.10.3 Android: How to Block a Number
- Go to your **Phone** icon.
- Click on the restricted call and then the ? symbol (may also say *Details).*
- Choose **Block Number** at the bottom of your screen.[102]

9.10.3.1 Set up a Personal Telephone Number Code

Carrier	Instruction

AT&T	https://www.att.com/esupport/article.html#!/wireless/KM1051397?gsi=Ks1FJro
Sprint	https://www.sprint.com/en/support/solutions/account-and-billing/learn-more-about-your-account-pin.html
T-Mobile	https://support.t-mobile.com/docs/DOC-37477
Verizon	https://www.verizonwireless.com/support/account-pin-faqs/

9.10.3.2 Set up a Separate Forwarding Telephone Number

Platform	Technology	Privacy Advice
Apple	Google Voice	https://itunes.apple.com/us/app/google-voice/id318698524?mt=8
	My Sudo	https://mysudo.com
	Others	https://www.makeuseof.com/tag/5-apps-getting-temporary-burner-phone-number/
Android	Google Voice	https://play.google.com/store/apps/details?id=com.google.android.apps.googlevoice&hl=en_US
	Others	https://www.makeuseof.com/tag/5-apps-getting-temporary-burner-phone-number/

9.11 SECURING YOUR PERSONAL EMAIL ADDRESS

Create unique disposable email addresses for different online accounts. *It is also highly recommended that you create a separate email address when opting out of your Digital Exhaust.*

- This can be read about at the URL *https://www.digitaltrends.com/computing/best-sites-for-creating-a-disposable-email-address/* and URL *https://mashtips.com/disposable-email-services/.*

10 Wi-Fi, Bluetooth and Near Field Communication

Threat actors can compromise devices over public Wi-Fi, Bluetooth, and Near-Field Communications (NFC), a short-range wireless technology.[103] This puts personal and organizational data, credentials, and devices at risk.

- Devices include laptops, tablets, mobile, wearable, and others that can connect to public wireless technologies.
- The guidance throughout helps users understand the risks in using public wireless technologies and enable them to make calculated decisions about the level of risk they accept.
- At a minimum, it is recommended users disable Wi-Fi, Bluetooth, and NFC when not in use.[104]

10.1 WI-FI

There are two kinds of Wi-Fi networks: secured and unsecured.[105] Most Wi-Fi networks that are created for home and business uses are password-protected and encrypted.[106]

- However, most public Wi-Fi hotspots are set up strictly for convenience – not security.[107]
- An unsecured Wi-Fi network can be connected to within range and without any type of security feature like a password or login.[108]
- In contrast, a secured network requires a user to agree to legal terms, register an account, or type in a password before connecting to the network.[109]

10.1.1 Public Wi-Fi Recommendations

It is recommended that you DO NOT:
- Allow your Wi-Fi to auto-connect to networks.
- Log into any account via an app that has sensitive information. Go to the website instead and verify it uses HTTPS before logging in.
- Leave your Wi-Fi or Bluetooth on if you are not using them.
- Access websites that hold your sensitive information, such as such as financial or healthcare accounts.[110]
- Log onto a network that is not password protected.

It is recommended that you DO
- Disable file sharing
- Only visit sites using HTTPS.
- Log out of accounts when done using them.
- Use a VPN, like Norton Secure VPN, to make sure your public Wi-Fi connections are made private.

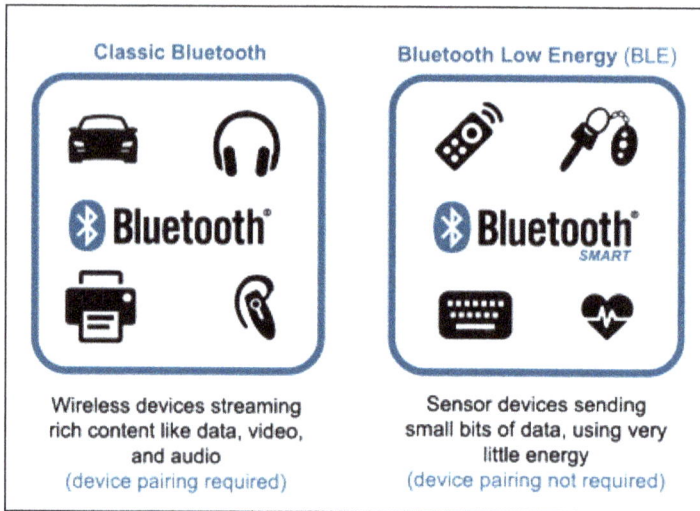

Classic Bluetooth — Wireless devices streaming rich content like data, video, and audio (device pairing required)

Bluetooth Low Energy (BLE) — Sensor devices sending small bits of data, using very little energy (device pairing not required)

10.2 BLUETOOTH

In the simplest terms, Bluetooth is the technology that enables exchange of data between devices within a short amount of distance.

- What separates Bluetooth radio waves from the broadcast sent out by a radio station is the fact that Bluetooth waves do not travel extremely far and are constantly switching frequencies.
- Most Bluetooth devices have a maximum connectivity range of about 30 feet, and that distance is reduced when obstacles are present.[111]
- Bluetooth Low Energy (BLE)—also known as Bluetooth Smart—is the latest version of Bluetooth technology that offers significantly less power consumption and costs compared to Classic Bluetooth while still supporting a similar communication range.[112]
- Bluetooth and Wi-Fi are often complementary, working at the same time and offering much the same connectivity, you may not always know which hardware is pairing with which devices.
- Just know that if in range, devices previously paired via Bluetooth will try to automatically connect.[113]

10.2.1 Bluetooth as an Attack Vector

There have been many noteworthy Bluetooth vulnerability discoveries in recent years and the sophistication of the attacks will only evolve.[114]

- Disturbingly, hackers no longer need to be nearby the devices to carry out their exploits.[115] Bluetooth was designed for short-range communications, but because they have radios, cyber thieves can exploit a system remotely and then use that system's Bluetooth interface to launch an attack.
- In this role, it is possible for an attacker to not only run these attacks remotely while in proximity, but also conduct them from much further away using low-cost equipment.

10.2.2 Notable Bluetooth Vulnerabilities

Because of an attackers' ability to implement remote attacks via radio, the increasing threat from Bluetooth devices to network security is a top concern for security teams. Here are the top eight recent Bluetooth vulnerability discoveries[116] that organizations have had to address:

10.2.2.1 BIAS (Bluetooth Impersonation AttackS)

Earlier this year, a new Bluetooth flaw dubbed BIAS was discovered with the potential to expose billions of devices to hackers. BIAS allows cyber-criminals to create an authenticated Bluetooth connection between two paired devices without needing a key.*[117]*

- The attacker can take over communication between the two devices by impersonating either end such as a mouse or a keyboard, giving the intruder inside access to the targeted device.*[118]*
- Once inside, the masquerading attacker can then implement malicious exploits such as stealing or corrupting data.[119]

10.2.2.2 BleedingBit

The attacker can use Bluetooth Low Energy (BLE) implementation vulnerabilities for remote code execution[120] and total machine take over to infiltrate networks[121].

10.2.2.3 BlueBorne

An attacker can actuate carefully constructed packets to cause buffer overflows[122], which can be exploited for code execution[123].

- The attackers can then take over a machine running Bluetooth Classic and use it as a potential entry point for malicious activity.[124]

10.2.2.4 Bluetooth Denial of Service (DoS) Via Inquiry Flood

This DoS attack targets BLE devices, running down their batteries and preventing them from answering other requests from legitimate devices.[125]

- This is particularly concerning for medical devices being used in life-saving situations.[126]

10.2.2.5 Fixed Coordinate Invalid Curve Attack

Hackers can crack the encryption key for both Bluetooth and BLE because of subtle flaws in the Elliptic Curve Diffie- Hellman key exchange process.[127]

- Attackers can imitate devices, inject commands, and penetrate for added security flaws.[128]

10.2.2.6 KNOB (Key Negotiation of Bluetooth)

An attacker can crack encryption on a Bluetooth conversation and then snoop to see all encrypted traffic as if it was plaintext.[129]

- The attacker can erase or inject packets, and ransom or publish the captured details.[130]

10.2.2.7 Malicious Applications Leveraging Radio Frequency Interfaces

Leveraging a downloaded app, a cybercriminal can access an iPhone's camera and microphone without permission.

- The attacker can then record and exfiltrate audio and video, and then ransom or publish the compromised information.[131]

10.2.2.8 Sweyntooth
An attacker within radio range can trigger deadlocks, crashes, and buffer overflows or completely detour security by sending faulty packets over the air.[132]

- If successful, this could result in the crash of devices such as medical equipment, potentially causing harm to patients, or other IoT connected devices in offices or homes.[133]

10.2.3 Bluetooth Beacons
If you own a business or are involved in marketing, you have some level of understanding about how beacon technology works[134] and you may have even received a Google beacon as part of Project Beacon[135], a program Google launched[136] to send free beacons to businesses with the aim of enabling proximity-based triggers and actions in both the digital and physical world. This Digital Exhaust is based on location-tracking data, gleaned from mobile phone users who have their Bluetooth enabled by default or by accident, as many people do.[137]

- With the emergence of COVID-19 in 2020, the issue of just how valuable and detailed our collective Digital Exhaust is has been proven by both Google[138] and Facebook[139] who began sharing location-tracking information with various authorities around the world to help them plan their COVID-19 containment strategies.
- The data supplied is "anonymized" and "aggregated", so there are no personally identifying markers. Nevertheless, the data does track people's movements - for example, Google's Mobility Reports[140], which it is made available for 131 countries and regions show foot traffic trends at various locations over time.

10.2.4 Securing Bluetooth
As a wireless data transfer standard, Bluetooth has some associated cybersecurity risks. You do not want unauthorized parties to access the data you are transferring via Bluetooth, nor do you want them to have access to your Bluetooth-enabled devices.

- It helps to know what the security risks with Bluetooth are so you can enjoy all the convenience of the widespread wireless technology while mitigating its risks.

10.2.4.1 Physically Secure Your Device
You may want to set up a "find my device" service on your phone through a trustworthy entity like Apple or Google so you have a way of using their technologies to find and remotely lock your device if you lose it.

10.2.4.2 Avoid Using Bluetooth to Communicate Sensitive Information
If you choose to use Bluetooth to transfer sensitive information from your device to another device, consider encrypting your files first.

10.2.4.3 Turning Off Bluetooth Discoverable Mode

- Ensure you turn off Bluetooth discoverable modes after pairing a new peripheral with your device.
- Once paired, you do not need to have discoverable mode on because your device will already know the peripheral's unique identifying code.
- This will also secure your device from any unwanted pairing attempts.

10.3 NEAR FIELD COMMUNICATION

Bluetooth and Wi-Fi while like near field communication on the surface, do have distinct differences.

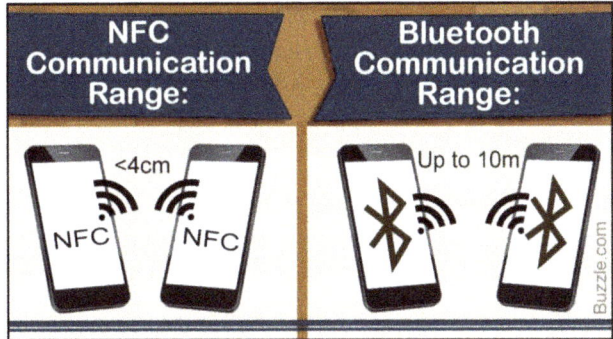

- All three allow wireless communication and data exchange between digital devices like smartphones.
- Yet near field, communication uses electromagnetic radio fields while technologies such as Bluetooth and Wi-Fi focus on radio transmissions instead.[141]

Near field communication, or NFC for short, is an offshoot of radio-frequency identification (RFID) with the exception that NFC is designed for use by devices within close proximity to each other.

- Devices using NFC may be active or passive. A passive device, such as an NFC tag, holds information that other devices can read but does not read any information itself. Think of a passive device as a sign on a wall. Others can read the information, but the sign itself does nothing except send the info to authorized devices.[142]
- Active devices can read information and send it. An active NFC device, like a smartphone, would not only be able to collect information from NFC tags, but it would also be able to exchange information with other compatible phones or devices and could even alter the information on the NFC tag if authorized to make such changes.

To ensure security, NFC often sets up a secure channel and uses encryption when sending sensitive information such as credit card numbers.

- Users can further protect their personal data by keeping anti-virus software on their smartphones and adding a password to the phone so a thief cannot use it if the smartphone is lost or stolen.[143]
- Unaccustomed users of near field communication, especially for payment purposes such as storing credit card information, may be concerned about the security and safety of their confidential information.

10.3.1 NFC Vulnerabilities

Security attacks include eavesdropping, data corruption or modification, interception attacks, and physical thefts. Below we cover the risks and how NFC technology works to prevent such vulnerabilities:

10.3.1.1 Eavesdropping

Eavesdropping is when a criminal "listens in" on an NFC transaction. The criminal does not need to pick up every single signal to gather confidential information. Two methods can prevent eavesdropping.

- First, there is the range of NFC itself.
- Since the devices must be close to send signals, the criminal has a limited range to work in for intercepting signals. Then there are secure channels.
- When a secure channel is set up, the information is encrypted and only an authorized device can decode it.
- NFC users should ensure the companies they do business with use secure channels.

10.3.1.2 Data Corruption and Manipulation

Data corruption and manipulation occur when a criminal manipulates the data being sent to a reader or interferes with the data being sent so it is corrupted and useless when it arrives.

- To prevent this, secure channels should be used for communication.
- Some NFC devices "listen" for data corruption attacks and prevent them before they have a chance to get up and running.

10.3.1.3 Interception Attacks

Like data manipulation, interception attacks take this type of digital crime one-step further. A person acts as a middleman between two NFC devices, receives, and alters the information as it passes between them. This type of attack is difficult and less common.

- To prevent it, devices should be in an active-passive pairing.
- This means one device receives info and the other sends it instead of both devices receiving and passing information.

10.3.1.4 Theft

No amount of encryption can protect a consumer from a stolen phone. If a smartphone is stolen, the thief could theoretically wave the phone over a card reader at a store to make a purchase.

- To avoid this, smartphone owners should be diligent about keeping tight security on their phones.
- By installing a password or other type of lock that appears when the smartphone screen is turned on, a thief may not be able to figure out the password and thus cannot access sensitive information on the phone.
- Through data encryption and secure channels, NFC technology can help consumers make purchases quickly while keeping their information safe at the safe time.

11 DEBIT AND CREDIT CARD TRACKING

Although it is illegal for financial institutions to sell your information, sharing your information is often important for their business operations and your information to be shared internally and with affiliates and non-affiliates.

- Affiliates are companies related by control or ownership, and non-affiliates are outside companies. The companies can be financial or non-financial in nature. Companies share your information with both parties to market to you.
- Some companies often claim a user's privacy would not be violated as all personal data has been de-identified and pseudonymized, (i.e., your personal information) like name and credit card number have been replaced by pseudonyms.
- If you would like to know more about privacy choices for your personal financial information, read the article by the Federal Trade Commission URL *https://www.consumer.ftc.gov/articles/0222-privacy-choices-your-personal-financial-information* and review the list of specific banks and credit card privacy opt-out links at *https://www.mymoneyblog.com/big-list-bank-credit-card-privacy-opt-out.html*

12 Social Media Platforms

The role of social media in our lives continues to grow each year and so too does the amount of personal information which can be found through our online personas.[144] [145]

- While who and what we share through social media is a personal choice[146], it is recommended that you be intentional about who you share your data with[147], to include which sites and platforms that you trust and consider worth the risk.[148]
- The role of the section below is to inform you of several privacy settings to aid you in securing your social network accounts so that you only share information with people you choose and not those you do not.
- Online social media services are teeming with private and public personal information.[149] [150]
- Control yours via the below links to privacy settings.
- Further, ensure your account usernames and/or account unique IDs **_do not_** correlate with your personal data, and do not respond to messages or accept connection requests from parties you do not know or cannot confirm to be legitimate.

12.1.1 Social Media Privacy Settings Links

Service	Privacy Settings
Facebook	https://www.facebook.com/about/basics
Instagram	https://help.instagram.com/196883487377501
Line	https://help.line.me/line/?contentId=20002865
LinkedIn	https://www.linkedin.com/help/linkedin/answer/92055/understanding-your-privacy-settings?lang=en
Pinterest	https://help.pinterest.com/en/article/edit-account-privacy
Skype	https://support.skype.com/en/faq/FA140/how-do-i-manage-my-privacy-settings-in-skype-for-windows-desktop
SnapChat	https://support.snapchat.com/en-US/a/privacy-settings2
Tumblr	https://tumblr.zendesk.com/hc/en-us/articles/115011611747-Privacy-options
Twitter	https://help.twitter.com/en/safety-and-security/how-to-make-twitter-private-and-public
Viber	https://support.viber.com/customer/en/portal/topics/592905-security-and-privacy/articles
WeChat	https://help.wechat.com/cgi-bin/newreadtemplate?t=help_center/topic_list&plat=2&lang=en&Channel=helpcenter&detail=1003386
WhatsApp	https://faq.whatsapp.com/en/android/23225461/?category=5245250
YouTube	https://support.google.com/youtube/answer/157177?co=GENIE.Platform%3DDesktop&hl=en

12.2 Facebook

Facebook is a social networking website where users can post comments, share photographs, and post links to news or other interesting content on the web, chat live, and watch short-form video. Shared content can be made publicly accessible, or it can be shared only among a select group of friends or family, or with a single person.[151]

Facebook's business model relies upon selling targeted advertising to you based on the personal information you share with it via its online social media services.[152] [153]The following techniques can help mitigate any personal risk you assume by using these services.

12.2.1 Standalone Email Addresses/Phone Numbers

Use a standalone email address that is not linked to any other account beyond Facebook. It is also recommended that you use a separate mobile number as well if possible.

12.2.2 Mobile Phone/Web Browser Settings

It is recommended that you ensure that your mobile phone and web browser privacy settings are properly configured.

- To ensure this, please go through and apply guidance on these topics elsewhere in this document. To do so please see Sections 3.7.1 and 3.7.2.

12.3 FACEBOOK ACCOUNT SETTINGS

12.3.1 Password Protection

Create a Facebook password different from the passwords you use to log into other accounts. For added tips, visit *fb.me/Passwords*. You can also test any sample password you choose at the URL *https://howsecureismypassword.net/*

12.3.2 Login Notifications

Facebook will send you a notification if someone tries logging into your account from a new device or browser.

- To learn more, visit *fb.me/LoginNotifications*

12.3.3 Login Approvals

Facebook will prompt you enter a special security code (*two-factor authentication*) each time you try to access your Facebook account from a new computer, phone, or browser.

- To learn how to turn on Login Approvals, visit *fb.me/LoginApprovals*.

12.3.4 Trusted Contacts

Trusted contacts are friends you can reach out to if you ever need help getting into your Facebook account.

- Once set up, if you are unable to access your account, your trusted contacts can access special, one-time security codes from Facebook via a URL.
- You can then call your friends to get the security codes and use those codes to access your account.
- To set up your trusted contacts, visit *fb.me/TrustedContacts*.

12.3.5 Login Location and Device Check

The **Where You are Logged In** section of your Security Settings shows you a list of browsers and devices that have been used to log in to your account recently.[154]

- You will also see the choice to End Activity and log yourself out on that computer, phone, or tablet.
- To review your active sessions and log out from unused browsers and apps, visit *fb.me/ActiveSessions*.

12.3.6 Customize Notifications

You can adjust what Facebook activity you are notified about and how you are notified.

- For more details, visit *fb.me/Notifications*.

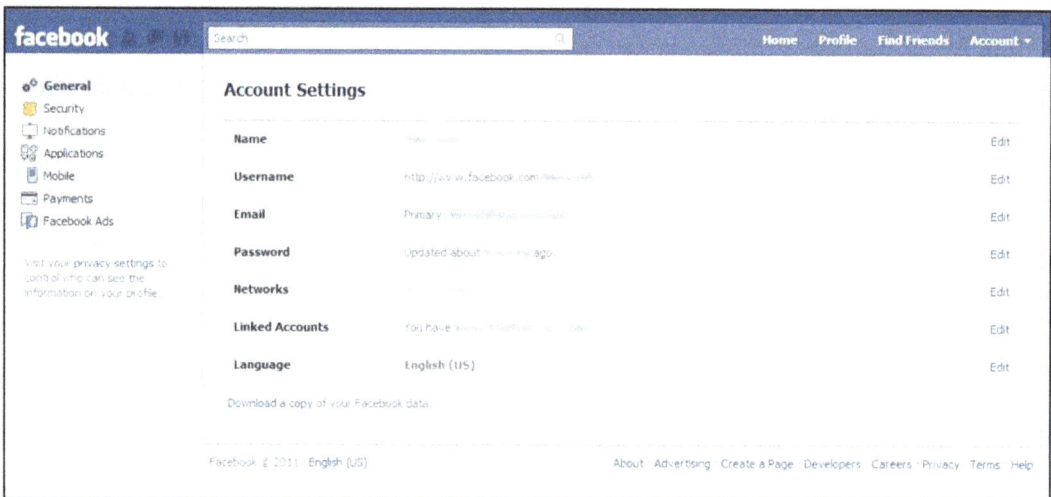

12.4 FACEBOOK SECURITY CHECKUP

Use Facebook's Security Checkup to review and add more security to your account.

- To start your own Facebook Security Checkup, visit *fb.me/securitycheckup*.

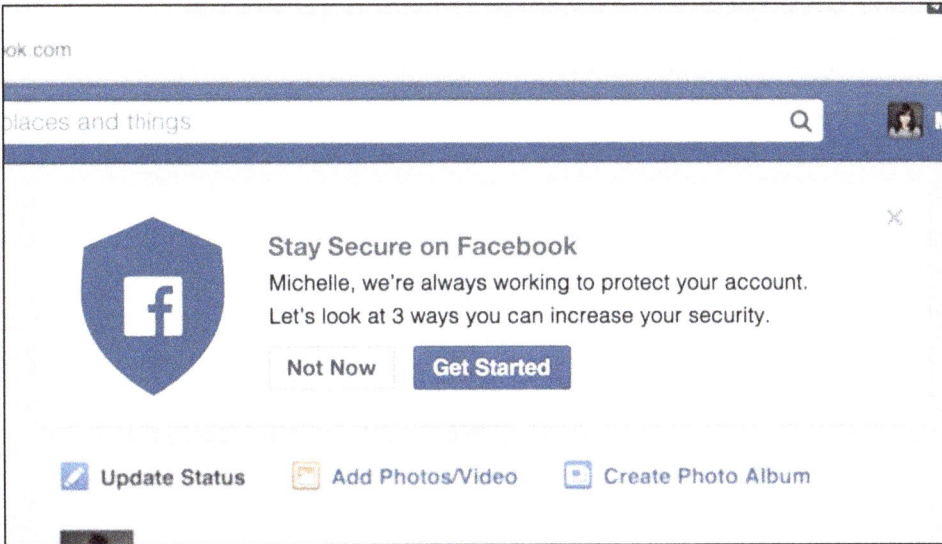

Figure 4. Facebook Security Checkup

12.5 FACEBOOK PRIVACY SETTINGS

12.5.1 Select Your Audience

Whenever you update your status, share photos, or post any information on Facebook, you can select who sees what you share through the audience selector tool.[155]

- This tool allows you to decide who sees what you share.
- The Custom option can be used to be as specific as you want for who can and cannot see something.[156]
- Facebook's help page will remind you *when you post to another person's Timeline, that person controls what audience can view the post*. *Additionally, anyone who is tagged in a post may see it, along with his or her friends.*
- To learn more about selecting audiences, visit *fb.me/AudienceSelector*.

12.5.2 Review and Approval

There are two options within the Timeline and Tagging Settings for reviewing content that is tagged.[157]

- The first choice allows you to approve or dismiss posts that you are tagged in before they appear on your Timeline.
- This automatically applies to posts where you are tagged by someone you are not friends with, but you can choose to review all tags by turning on the timeline review.
- The second choice allows you to approve or dismiss tags people add to your posts.
- When you turn this on, a tag someone adds to your post will not appear until you approve it.
- To learn how to enable tag reviews, visit *fb.me/TagReview*.

12.5.3 Search Engine Visibility

- If you do not want search engines to link to your profile, you can adjust your Privacy Settings.[158]
- However, some information from your profile can still appear in search engine results because it is information you shared to a Public audience or posts and comments you shared on Pages, Public groups, or the Community Forum section of the Help Center.
- To learn more, visit *fb.me/SearchEngines*.

12.5.4 Location Settings

Your location can be shared in many ways: with apps, by checking-in, via private messages, or by someone else tagging you.[159]

- It is important to consider when you share your location and with whom and to take measures to protect your location when possible.[160]
- To learn more about location privacy on Facebook, visit *fb.me/LocationPrivacy*.

12.5.5 View As Feature

You can see what your profile looks like to other people by using the View As tool.

- To learn more, visit *fb.me/ViewAs*.

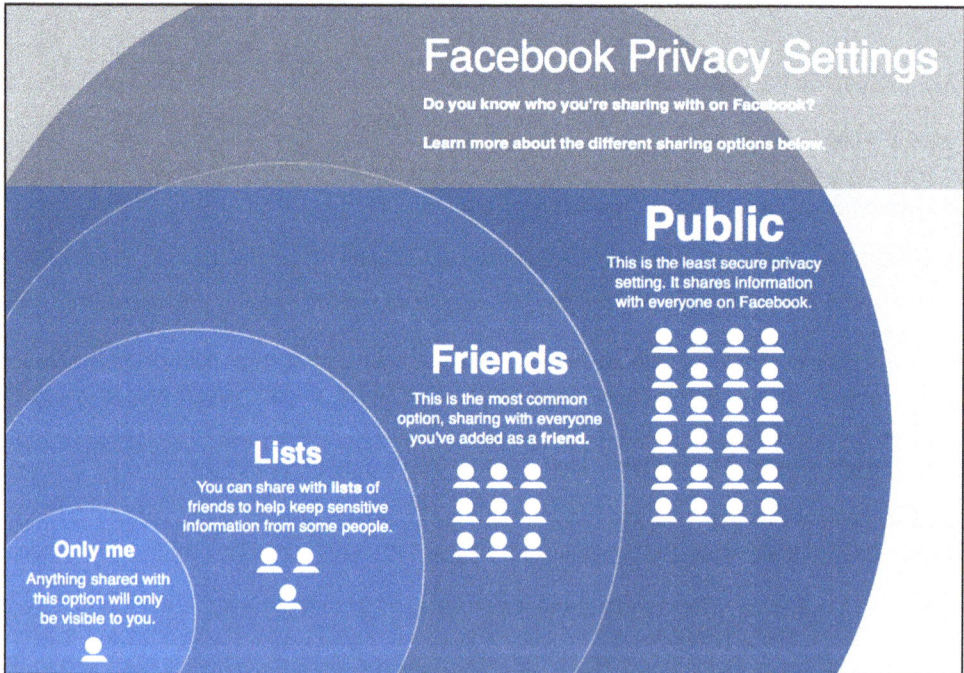

Figure 5. Facebook Privacy Settings

12.5.6 Disabling Advertising Features in Facebook

Go to your Account Settings and enter the section for Ad Preferences.

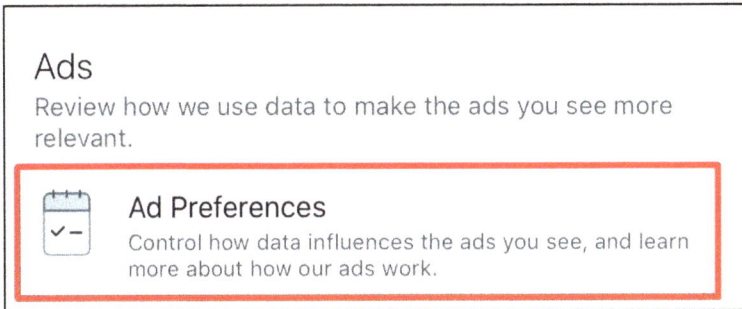

Then, enter each section **Advertisers**, **Your Information**, and **Ad Settings**.

12.5.6.1 Advertisers

Your Facebook account will have the same sub sections as highlighted below. They will educate you how Facebook already used your information for its advertising purposes.[161]

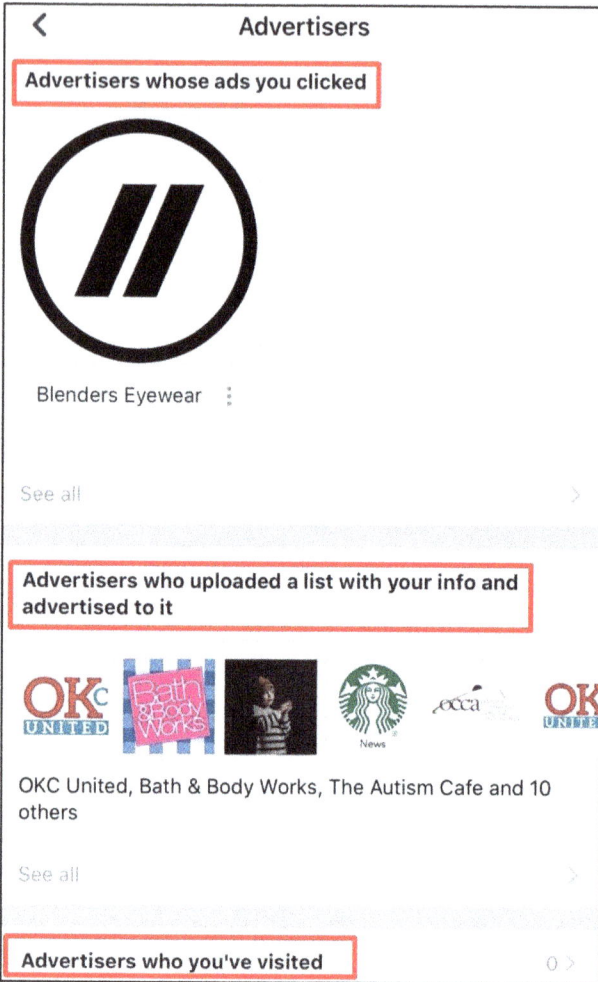

12.5.6.2 Your Information

Everything in this section is available to how Facebook serves advertising to you and your management of it **_does not_** affect how Facebook profile looks.

- Pay close addition to the **Review and Manage Your Categories** section; you may have Wi-Fi and Phone settings in it, which you can opt out of as well.

12.5.6.3 Ad Settings

Disable all Ad Settings under the sections entitled *Ads based on data from partners*, *Ads based on your activity on Facebook Company Products that you see elsewhere* and *Ads that include your social actions*.

Ads Settings

We use data to show better ads. You can use these settings to choose whether you want certain types of your data to influence the ads we show. Changing these settings won't affect the number of ads you see.

Ads based on data from partners

To decide which ads we show you, we use data that advertisers, app developers and publishers provide us about your activity off Facebook Company Products. This includes your use of partners' websites and apps and certain offline interactions with them, like purchases.

Not Allowed

Ads based on your activity on Facebook Company Products that you see elsewhere

When we show you ads off Facebook Company Products, such as on the websites, apps and devices that use our advertising services, we use data about your activity on Facebook Company Products to make them more relevant.

Not Allowed

Ads that include your social actions

We may include your social actions on ads, such as liking the Page that's running the ad. Who can see this info?

No One

12.5.7 Facebook Facial Recognition and Active Status

- Facebook describes facial recognition as *"Our technology analyzes the pixels in photos and videos, such as your profile picture and photos and videos that you have been tagged in, to calculate a unique number, which we call a template. We compare other photos and videos on Facebook to this template and if we find a match, we will recognize you. If you are untagged from a photo, or video, information from those untagged photos and videos is no longer used in the template. If your face recognition setting is set to off, we delete the template."* in their help center post at URL *https://www.facebook.com/help/122175507864081*
- Disabling active status allows you to run on the service private from other users and Facebook friends.

Privacy

Control who sees what you do on Facebook, and how data helps us personalize experiences.

Privacy Settings
Control who can see your posts and content, as well as who can search for you.

Face Recognition
Choose whether we recognize you in photos and videos.

Timeline and Tagging
Decide who can interact with you and your posts on Facebook.

Public Posts
Manage who can follow you, and who can comment on your public posts.

Blocking
Review people you've previously blocked.

Location
Manage your location settings.

Active Status
Show when you're active.

12.6 MANAGING YOUR FACEBOOK COMMUNITY

12.6.1.1 Friend Requests

Facebook is where so many of us connect with people we know personally, like friends, family, classmates, and coworkers. Facebook is based on authentic identities, where people are who they are in the real world.

12.6.1.2 Do Not Use Your Full Name on Facebook

This is one of the fastest ways to get into someone's life so you might as well make it harder for someone to find you if they get a hold of your personal information or use Facebook to gauge your life even in new social circles.

- Unfortunately, as Facebook notes, some individuals use tactics such as impersonating a friend to gain access to personal information.
- If you receive a friend request from someone you are already friends with, ask if they sent the new request before accepting it.
- If they did not create it, report the impersonating profile to Facebook.
- If you want to meet new people through Facebook, try connecting with Pages and groups that interest you.
- You can also choose to limit who can see your friend list if you are worried about your friends and family being contacted by someone.
- To learn more about adding friends and friend requests, visit *fb.me/FriendRequests*.

12.6.1.3 Unfriending

To unfriend someone, go to that person's profile, hover over the Friends button at the top of their profile and select Unfriend.

- If you choose to unfriend someone, Facebook will not notify the person, but you will be removed from that person's friends list.
- If you want to be friends with this person again, you will need to send a new friend request.
- To learn more about removing friends, visit *fb.me/Unfriending*.

12.6.1.4 Blocking

- Blocking a person automatically unfriends them and blocks them so they can no longer see things you post on your profile, tag you, invite you to event or groups, start a conversation with you, or add you as a friend.[162]
- Blocking is reciprocal, so you also will not be able to do things like start a conversation with them or add them as a friend.
- When you block someone, Facebook does not let them know you have blocked them. To learn more, visit *fb.me/Blocking*.

12.6.1.5 Reporting

Any type of content can be reported to Facebook. Facebook's Community Standards explain what type of content and sharing is allowed on Facebook.

- When something is reported to Facebook, a global team reviews it and removes anything that violates these terms.
- To learn how to report and what happens when you click report, click here *fb.me/Reporting*.

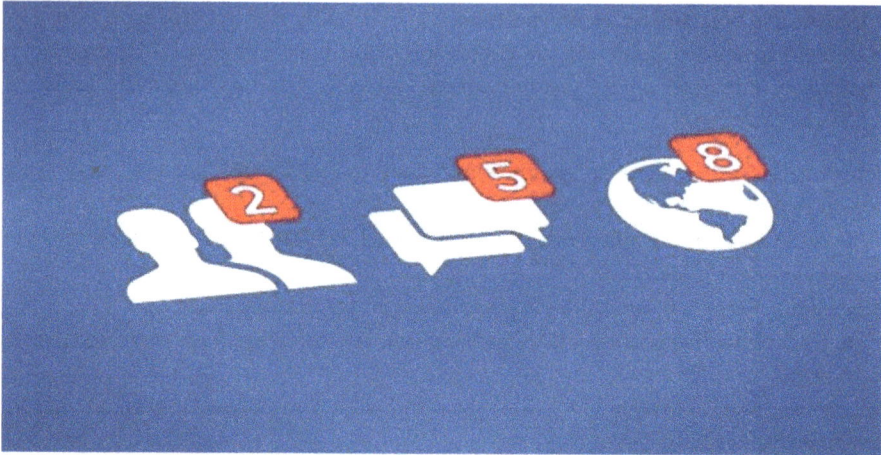

12.7 FACEBOOK MESSENGER

Facebook Messenger is a free messaging app and web-based platform that enables Facebook users to conduct instant message-based conversations with Facebook friends. Originally developed as Facebook Chat in 2008, the company updated the service and rebranded as Facebook Messenger in 2011.
Users of Facebook Messenger can send messages and exchange photos, stickers, audio, and files, as well as react to other users' messages, interact with bots, and conduct voice or video calls.
While Messenger was once limited to Facebook users only, it now powers conversations within Facebook, Instagram, Portal, and Oculus VR.[163]

12.7.1 Disabling Facebook Messenger from Automatically Syncing Your Contacts

12.7.1.1 If You Are Installing the App

Pay close attention to what prompts appear on your Mobile Phone as you install Facebook Messenger. After you have installed the App, you will begin setting up your profile based on existing Facebook information or whatever information you have provided.

- You will then see a prompt on your screen with two animated creatures. If you read the dialogue carefully (as highlighted in the graphic), you will see the text, which shows **"Continuously uploading your contacts helps Facebook and Messenger suggest connections and provide and improve ads for you and others and offer a better service."**
- Make sure you click on **'Not Now'**.
- This will prevent Facebook Messenger from uploading your contacts into the Facebook ecosystem.

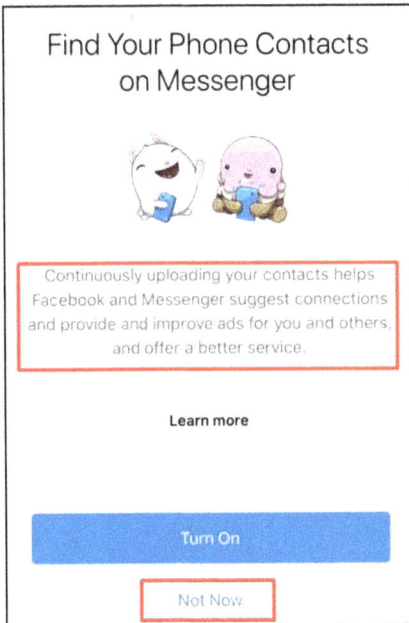

Find Your Phone Contacts on Messenger

Continuously uploading your contacts helps Facebook and Messenger suggest connections and provide and improve ads for you and others, and offer a better service.

Learn more

Turn On

Not Now

12.7.1.2 If the App Is Already Installed

You were unaware that enabling the feature discussed above actually uploaded your contact list from your Mobile Phone into the Facebook ecosystem so now you would like to go back, disable the setting, and now retroactively remove your contacts from Facebook Messenger.

- Here is how you disable the setting to stop continuously synchronizing your contacts with Facebook Messenger as well as remove them from Facebook's ecosystem.
- Launch the Facebook Messenger app from your Mobile Phone or Personal Device and go to the home screen.
- Look for the photo icon at the top left-hand corner of the screen and Tap on it.
- Now tap on '**People**' within the 'Preferences' section **(as highlighted in the graphic below).**
- Now tap on '**Upload Contacts**' and ensure you have the setting adjusted to '**Off**'.

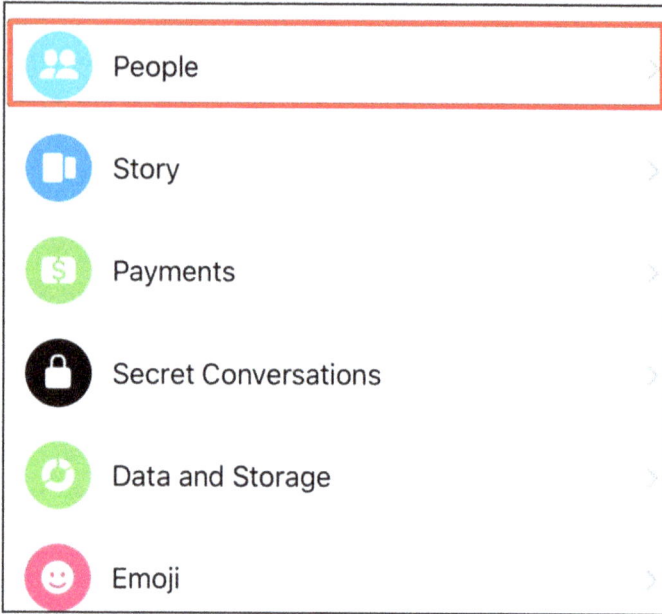

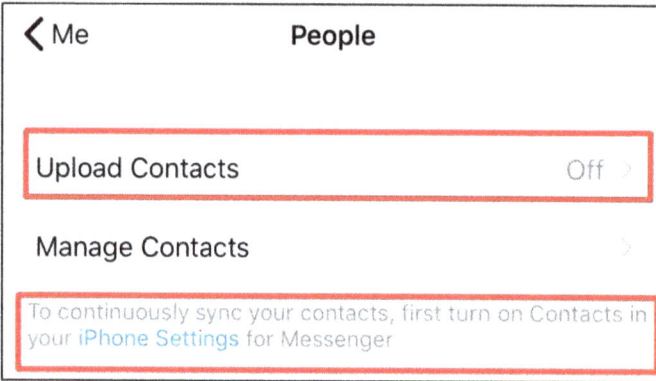

12.7.1.3 Stopping Facebook Messenger from Automatically Syncing Your Contacts (If the App Is Already Installed)

According to Facebook at URL *https://www.facebook.com/help/messenger-app/838237596230667* when you turn off contact uploading, the contacts you have uploaded to Messenger will automatically be removed.

- You can also go to the *Manage Your Uploaded Contacts* screen and tap **Delete All Contacts > Delete All Contacts** to delete these contacts. To stop your contacts from being uploaded again, you will need to turn off contact uploading on any devices where you are using the Messenger app.

Done Manage Your Uploaded Contact... ⬆️

Manage Your Uploaded Contacts and Call and Text History

These are the contacts and call and text history that you've uploaded from Messenger. Information like this helps Facebook and Messenger make better suggestions for you and others, and helps us provide a better service.

You may have uploaded info about these contacts beyond just the phone numbers below, like nicknames. You can see that data by visiting our Help Center.

To stop continuously uploading your contacts, turn off the Sync Contacts setting in the Messenger app. To stop continuously uploading your call and text history, turn off the Continuous Call and SMS Matching setting. Turning off each setting will delete all of your previously uploaded contacts or call and text history from Messenger.

Keep in mind that if you delete the information on this screen, but have continuous uploading still turned on for either setting, the info will be uploaded again automatically.

See contacts you've uploaded from Facebook.

| **Contacts** | **Call Logs** |

⊖ Delete All Contacts

12.7.2 Additional Facebook Messenger Privacy Settings

You can control your privacy in Messenger by choosing who can see your active status, choosing your Story audience, using secret conversations and more. Here are some ways to control your privacy in Messenger.

12.7.2.1 Control who can see when you are active

Active Status shows your friends and contacts when you are active or recently active on Facebook or Messenger.

- The following link will instruct you on how to *control your active status in Messenger*.

12.7.2.2 Control Chat Lists

If someone who you are not connected with on Facebook sends you a message, you will receive a connection request.

- The following link will instruct you on how to *control who can start a new chat with you in Messenger*.

12.7.2.3 Secret Conversations

Secret conversations in Messenger are end-to-end encrypted and can only be read on one device of the person you are communicating with. The following link will instruct you on how to use *secret conversations in Messenger*.

12.7.2.4 Clear Your Search History

Facebook Messenger allows users to edit or clear their search history in Messenger. The following link will instruct you on how to *clear your search history in Messenger*.

12.7.2.5 Remove Sent Messages

Facebook Messenger allows users to permanently remove a message that you have sent for everyone in the chat, or just for yourself.

- The following link will instruct you on *how to remove a message* within Facebook Messenger.

12.7.2.6 Customize Story View

You can control who can and cannot see your story.

- The following link will allow you to *choose who can see your story in Messenger*.

12.8 INSTAGRAM

Instagram is a free social networking service built around sharing photos and videos. It launched in October 2010 on iPhone first and became available on Android in April 2012. Facebook bought the service in April 2012 and has owned it since. Like most social media apps, Instagram allows you to follow users in which you are interested. This creates a feed on your homepage, showing recent posts from everyone you follow. You can like posts, comment on them, and share them with other people.[164]

12.8.1 Instagram Start Screen

The graphic of Instagram's start screen can be found at the following link.

12.8.2 Open the Camera

When you are on the home tab, you can tap the "camera" icon in the top left-hand corner to start adding photos and videos to your Instagram profile.

- NOTE: You will need to allow Instagram to access your camera and microphone before you can use this feature.

12.8.2.1 Direct Messages

The "paper airplane" icon in the top right from the home tab will get you access to your direct messages.

- Here you can view messages from people as well as create direct messages to send to your connections.

12.8.2.2 The Home Tab

This is the default view when you open the Instagram app. It is also, where the media, images, and stories from the people you are following will appear.

- From the home tab, you have access to add photos and/or videos to your feed, access your direct messages, search, connect and access your profile settings.

12.8.2.3 The Search Page

The magnifying glass will take you to the **"Search"** page.

- From here, you can search for accounts, keywords, hash tags, and topics simply by typing in the **"Search"** bar at the top of the screen.

12.8.2.4 The Camera Page

By clicking on this button, you will see your phone's camera popup. From here, you can either choose to add a photo or video from your camera roll (already on your camera) or choose to take a new one.

- NOTE: You will need to allow Instagram to access your camera and microphone before you can use this feature.

12.8.2.5 Account Activity

The "heart" icon will take you to your account activity page.

- This is where you can see comments, likes, shares, and follows for your account, as well as the people you are following.

12.8.2.6 Profile and Account Settings

You can access your own profile and account settings by tapping on the little icon that looks like a person.

- Once on this tab, you can choose to add latest photos and videos, edit your profile and more once again.
- While on this page, tapping the 'hamburger icon' in the top right will slide out more options where you can view your **"saved posts"**, or access the **"discover people"** functionality to connect with your friends from Facebook, or access Facebook directly.

12.9 INSTAGRAM'S PRIVACY AND SAFETY CENTER

If you need added help in understanding the wide-ranging settings Instagram offers you as a user for safety and reporting threatening activity, the following Instagram help center link is extremely informative.

12.9.1 Privacy Settings

The following privacy settings should be enabled to make you safer while you are using the platform as well as ways you can reduce your Digital Exhaust.

12.9.1.1 Private Profile

This is the most popular privacy setting and one you should enable right away. By default, Instagram accounts are public, meaning; anyone on Instagram can view your photos, like and comment on them.[165]

- Thankfully, Instagram gives you a choice to make your profile private. When you have a private profile, only your followers can see your published photos and stories.
- This setting does not change your viewing method, as you can still see other public profiles' photos and stories.
- To make your profile private, first open the Instagram app and go to the profile screen. Then tap on the three-dot icon at the top-right corner to open Settings in case of Android phones. On an iPhone, tap on the gear icon.
- Under **"Settings"**, tap on **"Private account"** and turn it on. You must also tap on Account privacy and enable the setting **"Private account"**.

- It is unknown why Instagram has kept the same setting in two places. Per Instagram, business profiles are not able to make their accounts private.
- If you want to make your business account private, first switch back to a personal account.

12.9.1.2 Removing Followers

When you make your Instagram profile private, there will be many people in your Followers list that you do not want there. Previously, you had to block such users, but Instagram has changed that setting now.

- It is not necessary for you to have a private profile to remove followers; you can do this even if you have a public profile. According to Instagram, removing specific followers will not let them know about being removed.
- To remove Instagram followers, go to your **"Profile"** and tap **"Followers"**. You will see the three-dot icon next to every follower.
- Tap on it for the follower you would like to remove and select **"Remove"** on the pop-up screen. If you would like added screenshots, the following link is helpful.

12.9.1.3 Turning Off Your Activity Status

In 2018, Instagram launched an Activity status feature. It shows the last time users were active on Instagram and with whom they had direct conversation. In addition to your activity, Instagram also introduced the online status indicator.[166]

- When a person is online, you will see a green dot next to their username in Direct Messages (DM). Per Instagram, here are the steps to turn them off.
- Go to your profile and tap the three-dot icon or the gear icon. Scroll down and tap on **"Activity status"**. On the next screen, disable **"Show activity status"**. This will turn off activity status and green dot both.

12.9.1.4 Blocking Comments

Sometimes when people do not like a picture or video that you posted, they resort to trolling you in the comments. Instagram gives you the choice to turn off their comments.

- You can do this for all posts from the general **"Settings"** and even for an individual post. Per Instagram, here is how you what you need to do to stop comments on all your Instagram posts.
- On your profile, tap on the three-dot icon to go to **"Settings"**. Under **"Settings"**, tap on **"Comment controls"**.
- Then you will get two options: **"Allow Comments from"** and **"Block Comments from"**. You can use the first choice to white filter the comments. Meaning, only the people that you add here will be able to comment on your posts.
- On the other hand, when you block people from commenting, everyone else except these users will be able to comment.
- To turn off comments for an individual post, open the post and tap the three-dot icon at the top-right corner.
- Select **"Turn off commenting"**.

- You can also enable the setting **"Hide offensive comments"** as well as the **"Manual filter"** option.
- If you need to report offensive or abusive behavior, Instagram provides you with instructions on how to do so at the following link.

12.9.1.5 Stopping Direct Messages (DM)

Everyone on Instagram can message you, whether they follow you or not. However, messages from people other than your followers are kept under a separate folder (Requests) in DM. While Instagram does not let you stop DMs for normal messages, you can restrict DMs for stories.[167]

- Instagram offers three settings for message replies in stories: **"Everyone"**, **"People you follow"**, and **"Off"**.
- Here is how to set it. Open Instagram Settings by tapping the three-dot icon (Android) and gear icon (iPhone) on the profile screen.
- Next, tap on **"Story controls"** and under **"Allow message replies"**, select the preferred option.

12.9.2 Instagram's Privacy Settings & Information Link

If you need added help in understanding the wide-ranging settings Instagram offers you as a user, the following Instagram help center link is extremely informative.[168]

12.9.3 Disable "Resharing Posts to Stories"

If you have a public profile, people can reshare your posts on their stories along with your username. While some people may not have an issue with it, I certainly do, so here are the steps Instagram provides you the to turn this feature off.

- Open your Instagram Settings, scroll down, tap "Resharing" to stories, and ensure you have disabled this setting.

12.9.4 Hide a Story

Instagram offers different privacy settings for posts and stories. While you cannot change the privacy of individual posts, you can customize the privacy of your stories, which will allow you to hide stories from specific followers.

- To do so, launch Instagram Settings and tap on **"Story Controls"**. Select the followers from whom you want to hide stories under the **"Hide story from"** option.
- A couple important Privacy tips for you on sharing Instagram stories, Private posts you share to social networks may be visible to the public depending on your privacy settings for those networks.
- Instagram offers an example at the following link that a post you share to Twitter that was set to private on Instagram may be visible to the people who can see your Twitter posts. This is a prime example of how your Digital Exhaust can pop up in ways you least expect it.

12.9.5 Approve Tagged Posts

Instagram has a separate section for tagged photos and videos. When a person tags you, it will automatically be added to your profile. I think many of us have experienced situations where we have been tagged in pictures that are not good always. Therefore, it is better to approve tagged posts first. Once you approve them, only then they will be added to your profile.

- To enable this setting, continue to Instagram Settings and tap on **"Photos of you"**. From here you can disable the setting **"Add Automatically"**. If you would like to hide a photo or video you have been tagged in, the following link from Instagram will provide you steps to do so.

12.9.6   Clear Instagram's Search History
If you often search for a person or a hashtag, it will appear under the search tab in Instagram.

- To clear your search history, open Instagram Settings and tap on **"Search history"**.
- Then on the next screen, tap on **"Clear search history"**.
- If you have trouble cleaning out your search history, the following link is filled with steps should you want to go nuclear and really scrub data out.

12.9.7   Photo Metadata
Perceptive threat actors can exploit the start of each photo presents unique Digital Exhaust which when left unchecked. Regardless of whether loopholes exist within Instagram to exploit my photo metadata, I go to the trouble of removing all my EXIF data from my photos because I never know where my personal data will end up, particularly in cloud-based storage environments.

- It is recommended that you remove any EXIF data so you do not hand it to a third party should a data breach occur even if it is stripped from social media platforms or in texting exchanges.
- In addition, it is recommended that you turn off geotagging by default.
    - NOTE: When you turn off geotagging, it only applies to photos taken after you have turned off the location feature.

12.9.8   Location Data
I highly recommend NOT showing your location when posting.

- If you do not understand how Instagram's Location Tags work, the following link is extremely informative.
- If you need a hand locking own your Location data, check out https://help.instagram.com/519522125107875 which outlines how your personal device(s) collect and track your daily location and ways you can increase your awareness of this issue with all Apps or Devices you use.

12.9.9   Syncing Contacts and Finding People to Follow
When it comes to synchronizing your contacts from your Mobile Device to Instagram, I would HIGHLY DISCOURAGE you from doing so. As Instagram is part of the Facebook ecosystem, I have already covered the dangers of synchronizing your contacts.

- If you need added help understanding how Instagram works with syncing contacts and finding people, the following Instagram help center link is extremely informative https://help.instagram.com/519522125107875.

- Additionally, if you would like information on how to disconnect your Instagram account from another social network, the following Instagram help center link is helpful.

12.9.10 Resources for Parents
The following link https://about.instagram.com/community/parents will be immensely helpful for parents of children who use Instagram.

- Instagram has a simple interface that is easy for unaccustomed users to understand intuitively, no matter their age, there are several Privacy settings that are highly recommended a user enable.

12.10 LINKEDIN
LinkedIn is the world's largest professional network on the internet. You can use LinkedIn to find the right job or internship, connect and strengthen professional relationships, and learn the skills you need to succeed in your career. You can access LinkedIn from a desktop, LinkedIn mobile app, mobile web experience, or the LinkedIn Lite Android mobile app.[169]

12.10.1 Understanding Social Engineering on LinkedIn

12.10.1.1 Detecting Fake LinkedIn Accounts/Personas
This section will give tips for how to spot fake or "doppelgänger" LinkedIn accounts. This is critical because connecting with a fake LinkedIn profile can give cyber criminals or Advanced Persistent Threat actors access to important and powerful information about you, such as details about your history, company, and professional contacts.[170]

- That information can be used to create detailed and believable phishing campaigns and other financial swindles.
- In short, beware of LinkedIn accounts with fake photos, incomplete profiles, limited connections, fake names, poor spelling, and grammar, and/or suspicious work history.

12.10.1.2 Fake Photos
Model-quality photos often go with many Fake LinkedIn profiles.

- If you are suspicious about a photo, there is a straightforward way to check its authenticity. Simply do a reverse image search using **TinEye**, **Bing's Visual Search** or **Google's Reverse Image Search**.
- These search engines will show you where, if any place, the same image has been used previously online.

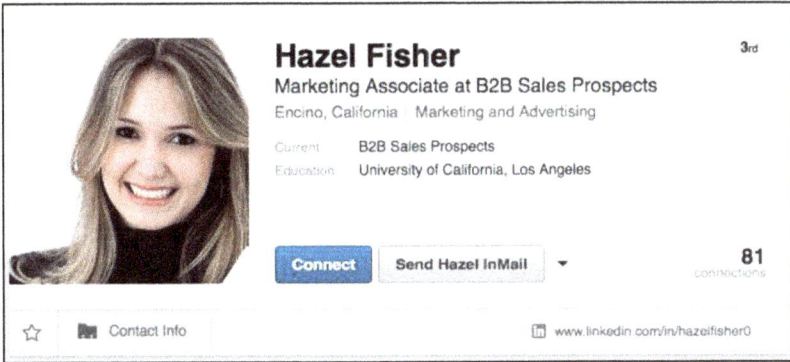

12.10.1.3 Incomplete Profiles

One key indicator of fake LinkedIn accounts is the lack of any information about the individual. If there is information, it is often in the form of mostly generic statements that lack any specificity in the summary and experience sections.

- Conversely, genuine profiles belonging to real people typically include a mixture of personal details, such as causes, volunteering, hobbies, education, recommendations, and the use of the first person when writing the 'Summary' or 'Experience' sections.
- Many fake profiles used for swindles do not bother to add personal information and keep detail to a minimum.
- Most people also personalize their custom LinkedIn URL while false accounts will not as they are created quickly and without tremendous attention to detail.
- This may not be the case for more sophisticated Cyber criminals or Advanced Persistent Threat actors.

12.10.1.4 Limited Connections

Genuine profiles typically have a mixture of people and profiles among its connections.

- Fake profiles may have connections with all the same or all opposite gender people with fake-looking profile pictures.
- Fake profiles can range from a few to several hundred connections, as well as a handful of skill endorsements.
- They also usually belong to several groups and follow a couple of companies and influencers.
- Check out mutual connections from a connection request, or better yet, message your connections directly to see if you can confirm an individual's identity prior to connecting to their profile.

12.10.1.5 Fake Names or Doppelgangers

Threat actors may create fake names or doppelgänger accounts to help their threat activities.

- Accounts created in this may use generic names or that of a famous person, like an actor, actress or television personality.

- Some scammers will use the name of a more obscure actor or actress that would not be as known to most of those on LinkedIn.
- Threat actors may also create accounts that impersonate a legitimate person's account. These accounts are doppelgangers, and their users try to assume a legitimate connection's identity as best as they can.
- These doppelganger accounts are often 3rd degree connections. To protect against this, run the account name in LinkedIn's search function to see if they have more than one account.
- If so, you may have showed their doppelgänger or found the true account and uncovered that whomever you are interacting with is the doppelganger.
- If you can, block the illegitimate account(s). This prevents the threat actor from viewing your profile, trying to follow your account on LinkedIn, and from delivering any type of malware to you through LinkedIn InMail.

More about doppelgänger accounts are available in the article "A Sneak Into The Devil's Colony-Fake Profiles in Online Social Networks" at URL https://arxiv.org/ftp/arxiv/papers/1705/1705.09929.pdf

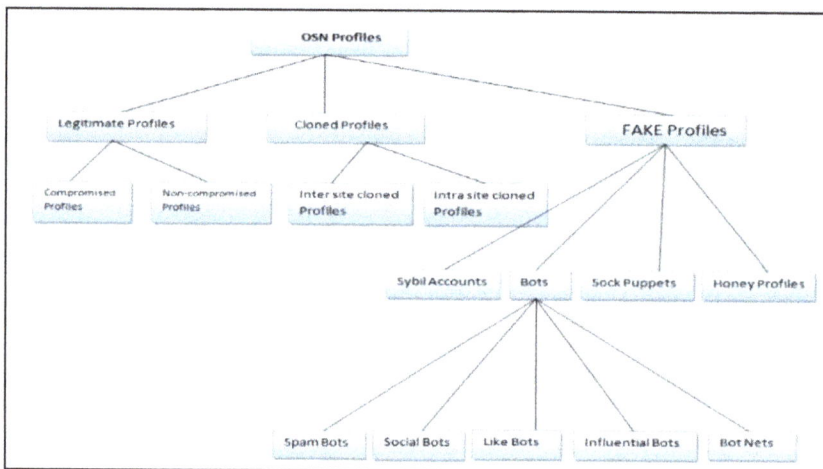

12.10.1.6 Poor Spelling and Grammar
Many fake profiles include obvious errors like misspellings and poor grammar. Often, the first name is displayed in all capital or lowercase letters, which would not be common to see in a genuine profile.

12.10.1.7 Suspicious Work History
One of the most effective ways to detect a suspicious work history is to check a connection's work experience by looking for their current employer elsewhere online and see if the person with the suspect profile is, in fact, listed as working there.

12.10.1.8 Suspicious Connection Requests
Be sure to vet connection requests if they have content with languages unfamiliar to you. Use the *Google Translate App* at URL *https://translate.google.com/intl/en/about/* if you want to read what the profile says in any unfamiliar language. On a mobile phone, take a screen shot and import it.

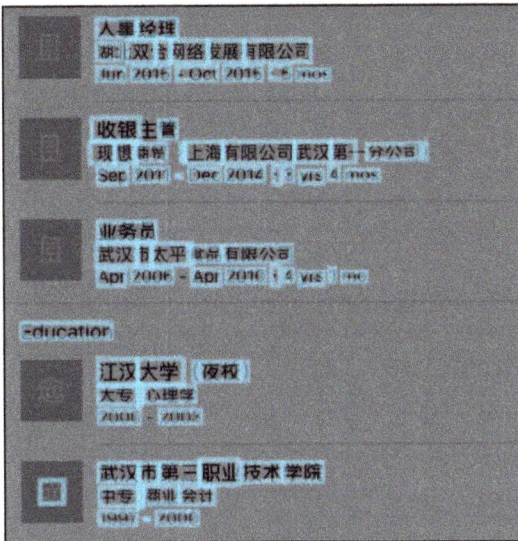

12.10.2 LinkedIn Privacy Settings

LinkedIn provides users with several privacy options.

- Review the following URL to better understand **them**
  https://www.linkedin.com/help/linkedin/answer/92055/understanding-your-privacy-settings?lang=en then head over to begin controlling them.
- You can control them at the URL *https://www.linkedin.com/help/linkedin/answer/66*

12.10.3 Settings & Privacy Page

The Settings & Privacy Page is organized into four tabs to help you easily view and change your account information, privacy preferences, ads settings, and communication notifications to include:

- Account tab - allows you to manage your account settings, such as adding email addresses, changing your password or language, and other account management options.
- Privacy tab - covers all privacy and security settings related to what can be seen about you, how information can be used, and downloading your data.
- Ads tab - enables you to control the information that LinkedIn uses to show you relevant ads by adjusting your account's ads settings.
- Communication tab - houses your preferences for how LinkedIn and other parties can contact you, and how often you would like to hear from us.

12.10.4 Linked In Account Settings

You can also check out the following information to learn more about some key settings you can manage through the Settings & Privacy page to include:

- Changing Your Password

- [Adding or Changing Email Addresses](#)
- [Adding and Removing Mobile Phone Numbers from Your Account](#)
- [Stopping or Changing Email Notifications](#)
- [Sharing Profile Changes with Your Network](#)
- ["Who's Viewed Your Profile" - Overview and Privacy](#)
- [Turning on Two-step Verification for Improved Security](#)
- [Setting push notification settings](#)
- [Viewing your groups](#)

12.10.4.1 Profile Photos on LinkedIn

You can suppress your profile photo from being displayed to everyone and only to people you confirm.

Choose whether to show or hide profile photos
of other members

Select whose photos you would like to see.

No one

✓   Your connections

Your network

All LinkedIn members

12.10.4.2 How Your Name Appears on Your Profile

LinkedIn allows you to control how people see your last name on the platform. Hide your last name from people not connected to your account.

12.10.4.3 Reviewing Where Your Name Appears on Your Profile

Modify your account's custom URL on your LinkedIn profile to omit your full name.

- In addition, it is recommended you do not openly post your resume online.
- It is also advised that you review any recommendations you receive and ensure your last name is controlled on them and any other personally identifiable information is not visible in them.

12.10.4.4 Follow Accounts Instead of Connecting to Them

According to LinkedIn, *"Connections are members who connected on LinkedIn because they know and trust each other. If you are connected to someone, you will both be able to see each other's shares and updates on your LinkedIn homepages. You can also send messages to your connections on LinkedIn. **Following someone on LinkedIn allows you to see the person's posts and articles on your homepage without being connected to them.** However, the person you're following won't see your posts."*

- More is available at the URL https://www.linkedin.com/help/linkedin/answer/32504/similarities-and-differences-between-following-and-connecting?lang=en
- The Following feature is a valuable tool provided by LinkedIn. It enables sensitive and high profile users to overtly control to whom their accounts connect. Users can always view a list of your followers on your profile page at URL https://www.linkedin.com/help/linkedin/answer/2717 and manage who can follow their updates at URL https://www.linkedin.com/help/linkedin/answer/53652
- Do this to ensure no suspicious or nefarious individuals are remotely viewing your LinkedIn profile.

12.10.4.5 Searching for People on LinkedIn

Assuming you controlled your account's last name and photo, it is more difficult for threat actors to spoof your LinkedIn account with a doppelgänger account.[171]

- Regardless, search your name in LinkedIn to look for any 3rd-degree connections who may be trying pass themselves off as the real you.
- The article at URL ***https://www.linkedin.com/help/linkedin/topics/6001/6008/3544*** offers a great overview for how to search.
- You can also perform Boolean searches on LinkedIn. Instructions for how to do this are available at URL https://www.linkedin.com/help/linkedin/answer/75814/using-boolean-search-on-linkedin?lang=en
- Also, if you want a better understanding about how your network and degrees of connection work on LinkedIn, read the article at URL https://www.linkedin.com/help/linkedin/answer/110/your-network-and-degrees-of-connection?lang=en

12.11 SNAPCHAT

Snapchat markets itself as a social media platform on which sent images and messages are only available for a limited amount of time.[172] The time limit is set by each individual user.[173]

- A recipient, however, can still take a screenshot of sent photos or chats or use another device to take photos of any sent material (users are notified when their message has been screenshotted).
- Further, there are many other ways in which people can collect information about a Snapchat user particularly if that user does nothing to change their privacy settings.[174]

12.11.1   Start Screen

Opening the mobile Snapchat app immediately opens that device's camera. To navigate to other pages of the application either select another choice at the bottom of the screen or the yellow silhouette in the top left-hand corner to navigate to your profile page. The profile page looks like this:

- First, make sure that simply because someone has your phone number or email, they cannot search for you using that information on Snapchat. Instead, they would need your exact username. Selecting the gear icon in the top right corner of your profile page will navigate you to the settings page.
- Select Mobile Number then uncheck **'Let others find me by using my mobile number'**. Repeat the process for email. Now if a blocked caller tries to find you via Snapchat, it will be much more difficult:

12.11.2 Profile and Settings

This area allows users to access a variety of features to include using the two-factor authentication feature, turning off your location, managing target ads, controlling who contacts you, managing Snapchat's use of your contacts, and finally controlling who you share with.[175]

- To navigate to settings, go to the gear icon in the top right-hand corner of your profile page.

12.11.3 Enabling Two-Factor Authentication

This feature means that when logging into Snapchat, users must enter an added code (sent via SMS) after the password.

- Someone would need to have both your password and your phone to access your account.

12.11.4 Location Sharing

To turn off your location, control who contacts you, and control who you share information with, navigate to Settings and scroll to the **"WHO CAN"** section.

12.11.5 Ghost Mode

Select **'See My Location'** to turn on Ghost Mode (no one can see your location) or you can customize the location settings to allow certain users to see your location.

12.11.6 Contact Accessibility

Select **'Contact Me'** to make sure only your friends can contact you.

12.11.7 Information Visibility

Select **'View My Story'**, **'See My Location'**, and **'See me in Quick Add'** to control who can see your information.

12.11.8 Opting Out Of Targeted Ads

Go to Settings under the **"ADDITIONAL SERVICES"** section and select **'Manage'**.

12.11.9 Use of Contacts

When you first use the app, Snapchat asks if you would like to synchronize your contacts.[176]

- At this point, you can grant permission for the Snapchat app to access your contacts and make updates whenever you add a contact to your phone.
- If you originally allow Snapchat this access, you can change it later by unchecking **"Sync Contacts"** in your settings. Go to **'Manage'** under the **"ADDITIONAL SERVICES"** section and then select **'Permissions'**.
- Following the above recommendations can reduce a user's Digital Exhaust; however, following all these steps also reduces the usability of the app.
- Further, by not allowing Snapchat to synchronize with your contacts, you will have to manually search for someone in Snapchat to see if they have an account.

12.12 TikTok

TikTok (formally branded as musical.ly) is a freeware, cross-platform, short-form mobile video media application. TikTok uses a device's data plan or Wi-Fi to broadcast trending video media created by users. [177]

- The application is free to users and is supported by advertisements.
- TikTok users draw from a cadre of free tools to create content for sharing, as well as Livestream content that may use real-time filters.
- This application is used for mobile devices but also has workarounds for use in desktop computers.

12.12.1 TikTok Screen Management

TikTok supplies a Screen Time Management setting for a daily usage maximum (i.e., 40, 60, 90, or 120 minutes per day) that allows users the ability to pre-decide the daily time spent in the application.

- When the selected time is met, a password is needed to continue to use TikTok –presuming that a parent or guardian selects the required password or that the user will self-monitor the time limit.

- If you wish to limit time on the app, go to the **Digital Wellbeing** section of the Settings & Privacy page and use the "**Screen Time Management**" option to select your time limit.
- You can also set a pin code which will be used for both Screen Time Management and Restricted modes.

12.12.2 **Making your account Private**

- Launch the TikTok app
- Open the "**Me**" tab in the bottom right
- Next, tap the three vertical dots in the upper right
- Tap "**Privacy and Safety**"
- Tap "**Private Account**"; if your profile is in Pro Account, you need to switch to a personal account to make your profile private.
- Turn off "**Suggest your account to others**"

12.12.3 Turning off "Suggest your account to other"

By default, TikTok will share your content by featuring it on the **"For you"** pages of people you do not know.

- If you want to prevent strangers from seeing your videos, you can turn off the **"Suggest Your Account"** choice.
- Turning this setting off will stop your account being recommended to other users and prevent other people from finding the account via search engines.

12.12.4 **Making Videos Private**

TikTok allows you configure previously posted or latest videos with specific privacy settings. Videos previously posted can be configured as follows:

- Open a video
- Tap the three-dot icon at the bottom right
- Select Privacy settings
- Tap "**Who can view this video**"
- Select **Friends** or **Private**

Newer videos can be configured as follows:

- Before uploading, tap "**Who can view this video**"
- Select **Friends** or **Private**

12.12.5 **Managing Duet Control**

You can control who can duet on your videos, which can be configured as follows:

- Go to the **"Privacy and safety"** settings choice under the app settings
- Tap '**who can duet with your videos**'
- Choose '**Friends or No one**' to limit those who can duet with you or your child

- You can do this for several different options such as who can send you direct messages and download your videos

12.12.6 Blocking Interactions

TikTok users can interact with your account and content in multiple ways: they can view or download it, direct message you, and duet with your videos.

- The default setting for these interactions is "**On**," but you have the choice to change it to **Friends** or **Off**.
- To limit how other users can interact with your videos go to the Safety section of the Privacy page.
- Blocking interactions stops comments, duets, and reactions, and prevents people from seeing your messages or the videos you have liked.

12.12.7 Reporting a User

To block and/or report a user on TikTok you can do so through the following steps:

- Go to the user's profile and tap the three dots at the top of the screen
- From the options select '**Block**' or '**Report**'
- If you block the user, it will ask you to confirm this
- If you wish to simply report the user, you need to select why you are reporting them

12.12.8 Enable 2-Factor Authentication

It is always worth enabling 2-factor authentication to add a layer of extra security on you and your child's account. *The verification code can be sent to either your mobile phone or email address.*

- Select "**Security**" in the settings and privacy menu
- Tap on "**2-step verification**"
- Select your chosen verification method "**Phone**" or "**Email**"

12.12.9 Hacking Attempts and Security Alerts

TikTok has a built-in feature to aid in detecting hacking attempts and suspicious activity on your account.

- By accessing your security alerts, shown below, you can see what devices have accessed your accounts or are trying to access your account without you, you can see what devices have accessed your accounts or are trying to access your account without your permission.

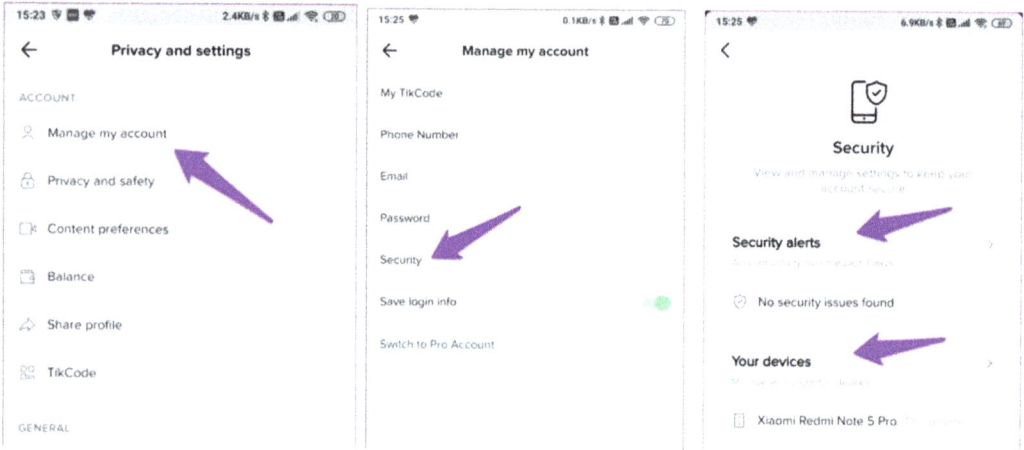

12.12.10   How to Download TikTok Data

Just like other social media platforms, TikTok also allows you to download your data.

- The option is available under "**Personalization and data**" under "**Privacy and safety**".[178]
- Tap on "**Download TikTok Data**", and under the '**Request a Data File**' tab, tap on the Request Data button to start the process.
- You will receive a confirmation email, followed by the actual file, which is usually sent within four days.
- The file will also be available under the "**Download Data**" tab. This file can be large, depending on how many videos have you uploaded, but that is not the only thing it will have.
- Your contact details and user activity, which includes comments and likes, are also included.

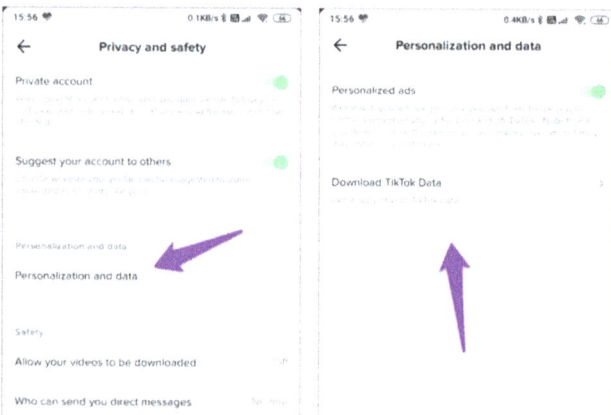

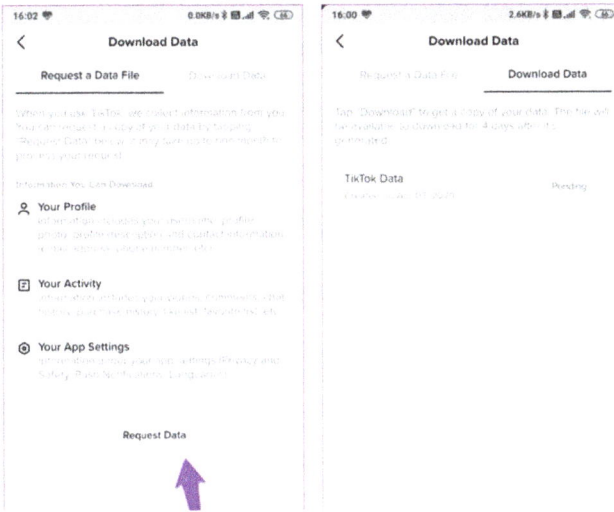

12.12.11   Digital Wellbeing Section: Child Safety - Family Safety Modes – Screen Time

12.12.11.1   Child Safety Settings
Restricted Mode stops most inappropriate content from appearing for children.

- It is also possible to set a passcode to prevent your child from changing this setting later. This setting is also found in the **"Digital Wellbeing"** section.

12.12.11.2   Family Safety Mode
This setting allows you to assign an account as **'Parent'** and **'Teen'**. This gives you remote access over an adolescent's TikTok account.

- Once connected to the account, you can control Screen Time Management, set how long your child can spend on TikTok each day.

12.12.11.2.1   Direct Messages
This feature allows you to control who can message your child or turn off direct messages completely.

12.12.11.2.2   Restricted Mode
This feature allows users to restrict types of content that you think are inappropriate for your child.

- It is possible to manage all this from a remote device, so you can make sure your child is always protected.
- This setting is also found in the **"Digital Wellbeing"** section.

12.12.11.3   Manage Screen Time:

If you wish to limit time on the app, go to the **"Digital Wellbeing"** section of the **"Settings & Privacy"** page and use the Screen Time Management option to select your time limit.

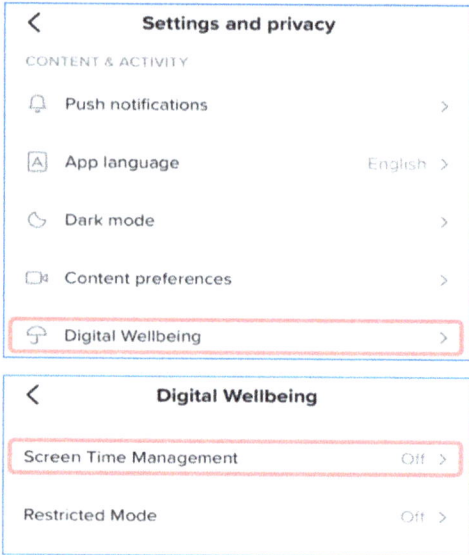

12.13 TWITTER

Twitter is an online news and social networking site where people communicate in short messages called tweets.[179] Twitter allows users to communicate and stay connected through the exchange of quick, frequent messages.[180]

People post Tweets, which may hold photos, videos, links, and text.[181] These messages are posted to your profile, sent to your followers, and are searchable on Twitter search.[182]

Twitter has extensive information on how to protect your account at *https://help.twitter.com/* [183]

- You can also find additional information on how to check safety and security settings as well as *How to protect your personal information.* [184]

12.13.1 Sharing Your Personal Information

When someone else Tweets your personal information such as in a doxing attack, you have the right to report the individual to Twitter.[185]

- However, if it is discovered that your personal information is publicly available, Twitter may not request that your information be removed.[186]
- Twitter also provides a link to assess ways to *protect your personal information*.

12.13.2 Your Profile
- In the Twitter menu, click **Profile**.
- Below your header photo, click **Edit profile**.
- This section will allow you to edit your **Bio**, **Location**, and **Website**. It should be noted that this information *will be displayed publicly* unless you adjust the privacy settings.
- Next to **Birth date**, you have the option to click **Edit**. Doing so will allow you to click **Remove birth date** to completely remove it from your profile.
- If you choose to display your birthday, you also have the choice to set the visibility for **Month and day** to something other than **Public** but leave the year as **Only you**.

12.13.3 Public Tweets versus Protected Tweets
When you sign up for Twitter, your Tweets are public by default, meaning anyone can view and interact with your Tweets.[187]

- Should you choose to protect your Tweets, you can do so through your *account settings*.
- Twitter provides extensive detail on how to configure settings for protecting your Tweets. If you protect your Tweets, you will receive a request when new people want to follow you, which you can approve or deny.
- Accounts that began following you before you protected your Tweets will still be able to view and interact with your protected Tweets unless you block them.
- Protected Tweets will not appear in third-party search engines and are only searchable on Twitter by you and your followers.

12.13.4 Photo Tagging
Even if your Tweets are protected, you can be tagged or mentioned in a photo.

- Likewise, your followers may re-share links to photos that you share in a protected Tweet.
- Links to photos shared on Twitter are not protected.
- Anyone with the link will be able to view the content.
- You can change who can tag you in a photo by visiting your Privacy and safety settings via twitter.com and Twitter for iOS or Twitter for Android apps.

12.13.5 Discoverability
Anyone with your email address or phone number can search for you on Twitter using this information.[188]

- In addition, anyone with this information in their contacts are provided your account (as a suggestion to follow) once they join Twitter.
- To turn this choice off, go to your privacy settings. Under Discoverability, uncheck **"Let others find you by your email"** and/or **"Let others find you by your phone"**.

12.13.6 Sharing Your Location in Tweets
Tweet location is off by default. You would need to opt in for this service.

- Once activated, Twitter will supply suggestions for locations of your next Tweet, but you can still choose not to share your location for individual Tweets.
- If you choose to enable precise location through Twitter's official apps, *this will allow Twitter to collect, store, and use your precise location*, such as GPS information.

12.13.7 Third-party Businesses and Personalized Ads

Even if you have turned off personalized ads and sharing data with third party businesses in your settings, Twitter shares information with business partners to help improve its business and ads will be shown based on your Twitter activity, information you have provided, as well as the devices you have used to log in.[189]

- Turning off these options simply reduces the relevance of the marketing activities on other sites, apps, and advertisements to you.

12.13.8 Blocking an Account

Blocked accounts cannot follow you, send direct messages to you, or tag you in a photo.

- They can view your public Tweets if not logged into Twitter.
- Blocked accounts do not receive a notification alerting them that their account has been blocked.
- However, if a blocked account visits the profile of an account, that has blocked them; they will see they have been blocked, unlike *mute*, which is invisible to muted accounts.

12.13.9 Two-Factor Authentication

Twitter offers two-factor authentication but instead of only entering a password to log in, you will also enter a code or use a security key.[190]

- This added step helps make sure that you, and only you, can access your Twitter account.
- During enrollment, Twitter will also verify that you have a confirmed email address associated with your account.
- After you enable this feature, Twitter will require your password, along with a secondary login method — either a code, a login confirmation via an app, or a physical security key to log in to your account.[191]

12.13.10   Account Access

- This feature allows you to *review the apps and devices connected to your Twitter account*.
- If there is any that do not truly need access to your Twitter account, click them, then click **Revoke access**.
- You can also access the **Sessions** section to review if there are any devices that do not truly need access to your Twitter account, click them, and then click **Log out the device shown**.

12.14 YouTube

YouTube is a video sharing service where users can watch, like, share, and comment and upload their own videos. The video service can be accessed on PCs, laptops, tablets and via mobile phones. Users of YouTube can search for and watch videos, create a personal YouTube channel, upload videos to their channel as well as like, comment or share other YouTube videos.[192]

12.14.1 YouTube Subscription Privacy Settings

- You can choose to make which channels you are subscribed to private or public.[193] By default, all settings are set to private.[194]

12.14.2 Public Listings

When your subscriptions are set to public, other users can see what channels you subscribe to.

- Your subscriptions are listed on your channel homepage. Your account is listed in the Subscribers List for any channel you subscribe to.

12.14.3 Private Listings

When your subscriptions are set to private, no other users can see what channels you subscribe to. Your account does not show in a channel's Subscribers List, even if you are subscribed.[195]

- If you take part in a subscriber-only live chat, other viewers will publicly see you are subscribed to the channel.

12.14.4 Privacy Channel Subscriptions

- Sign into YouTube.
- In the top right, click your profile picture.
- Click **Settings.**
- In the left Menu, select **Privacy**.
- Turn on or off **Keep all my subscriptions private**.

12.14.5 Hide Subscriber Count

By hiding your subscriber count, it will not be publicly visible to others on YouTube. You can still see your subscriber count from YouTube Studio.

- Sign into your Google Account.
- Go to YouTube Studio.
- Click **Settings** > **Channel** > **Advanced settings**.
- Under "Subscriber count," uncheck "Display the number of people subscribed to my channel."
- Click **Save**.

12.14.6 Location-based Recommendations

When you start using YouTube Music, location-based recommendations are turned off. Location helps YouTube Music offer you personalized music recommendations based on where you are. You can change your location-based settings to turn them on or off. Location history is automatically turned off for made for kid's content.[196]

- Visit music.youtube.com.
- Select your profile picture. 🖼
- Select Settings. ⚙
- Select Privacy.
- Make sure location-based recommendations are paused. This setting will prevent you from getting location-based recommendations.

12.14.7 Disable YouTube Ads

YouTube uses your data to improve your experience, like reminding you what you have watched, and giving you more relevant recommendations and search results.

- Your activity and information can also be used to personalize ads within YouTube and other Google Services. You can manage activity data in *Your Data in YouTube*.
- The ads that play on YouTube videos you watch are tailored to your interests. They are based on your Google Ad Settings, the videos you have watched, and whether you are signed in or not.
- You can control the ads that you see based on your Google Account *Ad Settings*. You can also *view, delete, or pause your YouTube watch history*.

12.14.8 Supervised Accounts for Kids on YouTube

Before you can begin setting up the supervised account for YouTube, you will need to have created your child's Google account through *Family Link*.[197]

- *Supervised YouTube accounts are available for kids under 13*; but that age may differ depending on what *country, you live in*.

Once this is done, you can begin setting up the supervised account for your child to explore YouTube. To do so, the following steps will walk you through that process.

- Open the **YouTube app** on your phone.
- Tap on your **profile picture*** in the upper right corner of the screen.
- Choose **Settings** at the bottom of the screen.
- Select **Parent Settings** towards the top of the page.
- If you have multiple child accounts created in Family Link, **choose the account you want to set up** for a supervised YouTube Account.
- Tap on **Set up YouTube**.
- Choose **SELECT** after **reviewing the information** about the type of content that may be available to your child.
- Pick the **content settings*** for your child's age.
- Scroll through the **Parent feature tour**, and then tap **NEXT**.
- Read the information **YouTube's privacy policies** and choose **FINISH SETUP**.

12.14.9 YouTube Kids Parental or Guardian Permission

You must be at least 13 years old to access YouTube Kids (*where available*) if enabled by a parent or legal guardian.[198]

- If you are under 18, you represent that you have your parent or guardian's permission to use the Service.
- It is recommended that your child read this agreement with you.
- You can find tools and resources to help you manage your family's experience on YouTube (***including how to enable a child under the age of 13 to use the Service and YouTube Kids***) in the *Help Center and through* Google's *Family Link*.

13 GOOGLE TRACKING AND LOCATION DATA

Google is an internet search engine. It uses a proprietary algorithm that is designed to retrieve and order search results to supply the most relevant and dependable sources of data possible.[199]

Settings are available to control Google's vast ability to collect data about you in its Activity Controls for your Google account.[200, 201, 202, 203] The easiest way to begin accessing the extensive controls that Google offers users is through the Google Safety Center found at URL *https://safety.google/privacy/privacy-controls/*

13.1 ACCOUNT PRIVACY CONTROLS

Browser Privacy Control	URL
Google Safety Center	https://safety.google/privacy/privacy-controls/
Google Account Privacy Checkup	https://myaccount.google.com/privacycheckup
Google Account Activity Controls	https://myaccount.google.com/activitycontrols
Google Dashboard (Manage All Of Your Google Data)	https://myaccount.google.com/dashboard
Control Web and App Activity	https://support.google.com/websearch/answer/54068?p=web_app_activity&authuser=0&hl=en&visit_id=637056287600533942-3442343815&rd=1
Manage Your Location History	https://support.google.com/websearch/answer/3118687?visit_id=637056287600533942-3442343815&p=location_history&hl=en&rd=1
Auto-Delete Web and App Activity	https://myactivity.google.com/myactivity?restrict=waa
Manage YouTube Privacy Settings	https://support.google.com/youtube/topic/9257518?hl=en
Your Google Data In Search	https://myactivity.google.com/privacyadvisor/search
Your Google Data In Maps	https://myaccount.google.com/yourdata/maps
Your Google Data In The Assistant	https://myaccount.google.com/yourdata/assistant
Download Your Google Account Data	https://takeout.google.com/settings/takeout?pli=1
Google Ad Settings	https://adsettings.google.com/authenticated?utm_source=udc&utm_medium=r
Google Maps Timeline	https://support.google.com/maps/answer/6258979
Search Activity	https://support.google.com/websearch/answer/54068?co=GENIE.Platform%3DDesktop&hl=en
Shared Usage and Diagnostic Data	https://support.google.com/accounts/answer/6078260
Google Security Tips	https://safety.google/security/security-tips/
Google Security Tips-Parental Supervision	https://safety.google/families/parental-supervision/
Google Security-Tips For Families	https://safety.google/families/families-tips

13.2 ASSISTANT DATA PRIVACY CONTROLS

In 2019, Google outlined substantial changes to how Google Assistant handles voice recordings.[204] These changes originated to meet users' expectations of data transparency.[205]

- If you use Google Assistant, the table below has the URL you can use to browse or delete your Google Assistant data to include your Web and App activity, Voice and Audio recordings, App and Contact information from your devices and Ad personalization.

Browser Privacy Control	URL
Google Assistant	https://myaccount.google.com/yourdata/assistant?e=PrivacyAdvisorAssistant&pli=1

13.3 CALENDAR PRIVACY CONTROLS

Gmail users are vulnerable to malicious or unsolicited Google Calendar notifications. Google Calendar allows anyone to schedule a meeting with you, and Gmail is built to integrate with this calendaring functionality.[206]

- When a calendar invitation is sent to a user, a pop-up notification appears on their smartphone.
- Threat actors can create messages to include a malicious link, which can be used in phishing schemes or social engineering attacks.[207]

Browser Privacy Control	URL
Google Calendar	1. https://support.google.com/calendar/answer/37083?hl=en
	2. https://support.google.com/calendar/answer/37082?hl=en&ref topic=3417970
Google Events	1. https://support.google.com/calendar/answer/6084018?co=GENI E.Platform%3DDesktop&hl=en

13.4 PRIVACY IN PERSONAL CONTENT

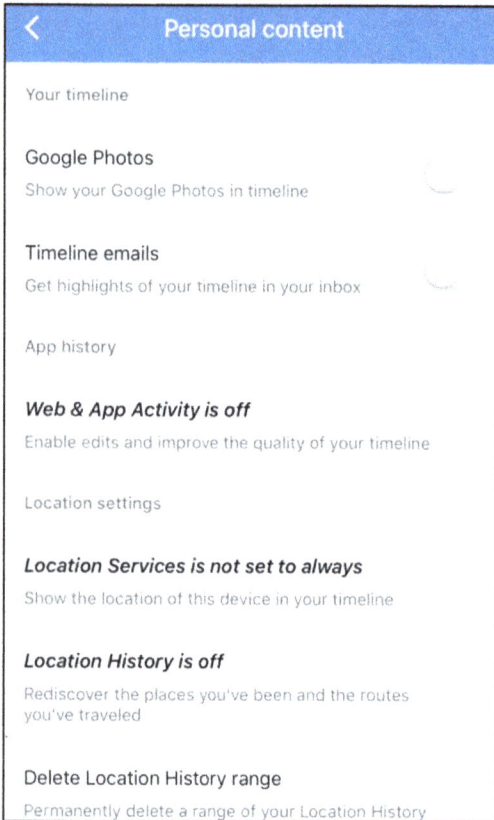

Personal content

Your timeline

Google Photos
Show your Google Photos in timeline

Timeline emails
Get highlights of your timeline in your inbox

App history

Web & App Activity is off
Enable edits and improve the quality of your timeline

Location settings

Location Services is not set to always
Show the location of this device in your timeline

Location History is off
Rediscover the places you've been and the routes you've traveled

Delete Location History range
Permanently delete a range of your Location History

Photo Library

Check for photos

This allows Google to periodically check for
photos you can add to places. Photos will never
be added without permission.

Profile settings

Show contributions on your profile

List all your reviews, photos, any public posts on
your public profile

Share profile with businesses

Make your public profile visible to businesses
you follow

Group similar faces
Manage preferences for face grouping   ∧

  Face grouping
  See photos of your favorite people grouped by similar faces. Learn more   ⬤

Sharing
Manage preferences for sharing   ∧

  Sharing suggestion notifications
  Receive notifications when you have new photos to share with friends   ⬤

  Remove video from motion photos
  Share only the still photos when sharing by link & in albums   ⬤

  Remove geo location in items shared by link
  Affects items shared by link but not by other means   ⬤

About, terms & privacy

Google Maps	© 2019 Google Inc.
Version	5.15.11
Terms of Service	
Privacy Policy	
Legal Notices	
Open source licenses	
Location data collection	Off
Clear application data	
Reset Google Usage ID	

14 AMAZON

Amazon is a cloud computing giant and the largest American e-commerce company.[208] Amazon collects your personal information with what you provide them[209] and will use your personal information to communicate with you about your purchases of products and services, improve and personalize your Amazon experience, and follow legal obligations, among others.[210]

- In addition, Amazon uses your personal information to display interest-based ads[211] for features, products, and services that might interest you and cookies and other identifiers to enable recognition of your browser or device.[212]

14.1 PRIVACY SETTINGS

Visit this link to learn about default Amazon settings to improve your privacy. Follow steps below to act at once. https://the-digital-reader.com/2019/04/11/six-default-amazon-security-settings-you-can-change-for-more-privacy/[213]

14.1.1.1 Removing Your Public Profile

Edit your name

This is how you'll appear to other customers.

Public name

Save

14.1.1.2 Private Shopping and Wish Lists

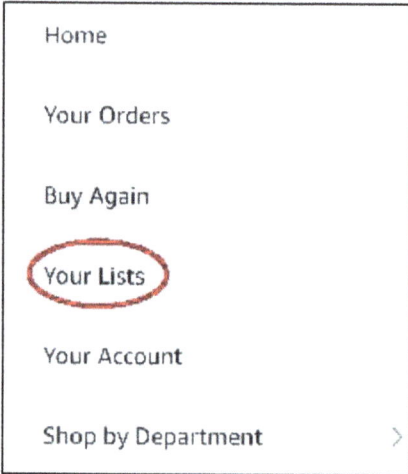

Then,

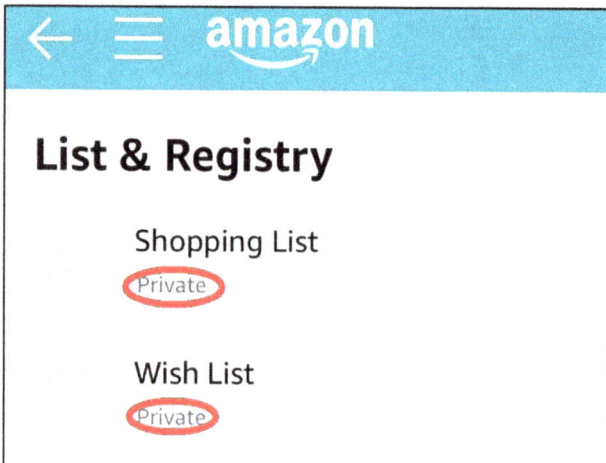

14.1.1.3 Browsing History and Tracking Cookies

Personalized content

Profile	>
Your uploaded product videos	>
Your Garage	>
Your Fanshop	>
Your Pets	>
Browsing history	>
Review your purchases	>

Then,

Manage History on This Device

Remove all items from view

Turn history on/off   Off

Turning off your recently viewed items will remove them from view.

14.1.1.4 Opting Out of Amazon Advertising Preferences

Skip this section if you would like Amazon's ability to track your activities and to market items to you.

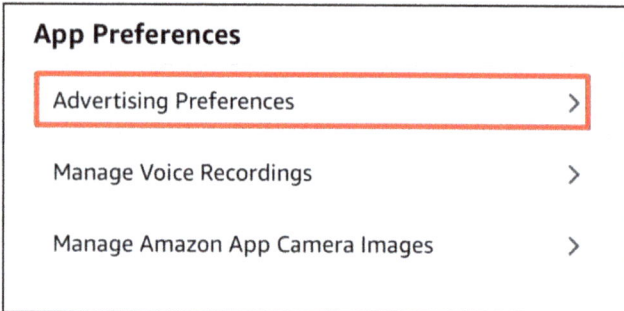

Then,

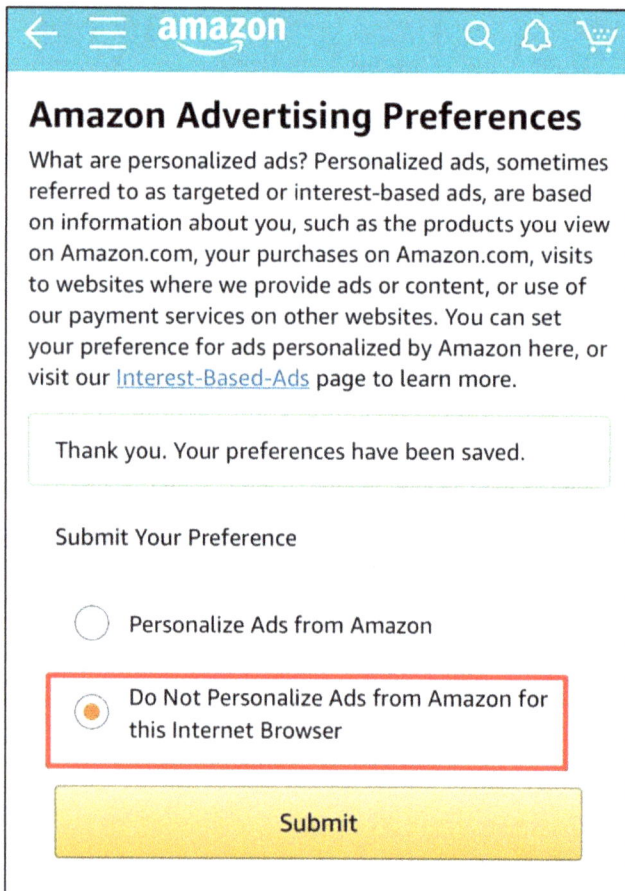

14.1.1.5 Disabling Amazon Saved Wi-Fi Passwords Feature

Ever wonder why you did not have to put your Wi-Fi password into your Fire TV or Alexa Echo? It is because this setting is enabled.[214]

Saved Wi-Fi Passwords ^

Your saved Wi-Fi passwords allow you to configure compatible devices so that you won't need to re-enter your Wi-Fi password on each device. Once saved to Amazon, your Wi-Fi passwords are sent over a secured connection and are stored in an encrypted file on an Amazon server. Amazon will only use your Wi-Fi passwords to connect your compatible devices and will not share them with any third party without your permission.
Learn more

Your Saved Wi-Fi Passwords
All Devices

Wi-Fi simple setup

Enable this setting to allow eligible devices to automatically use your saved Wi-Fi passwords during setup.

Wi-Fi simple setup is disabled | Enable |

14.1.1.6 Disabling Voice Recordings through the Amazon App

Manage Voice Recordings

When you use voice search with the Amazon App, we keep the voice recording associated with your account to learn how you speak to improve the accuracy of results provided to you and to improve our services.

You can choose to delete voice recordings you've made in the Amazon App that are associated with your account. This will delete these associated voice recordings you've made in the Amazon App on all mobile devices and may degrade your experience using voice features.

Delete Voice Recordings

✓ Your request was received

14.1.1.7 *Disabling Camera Images through the Amazon App*

App Preferences

Advertising Preferences ›

Manage Voice Recordings ›

Manage Amazon App Camera Images ›

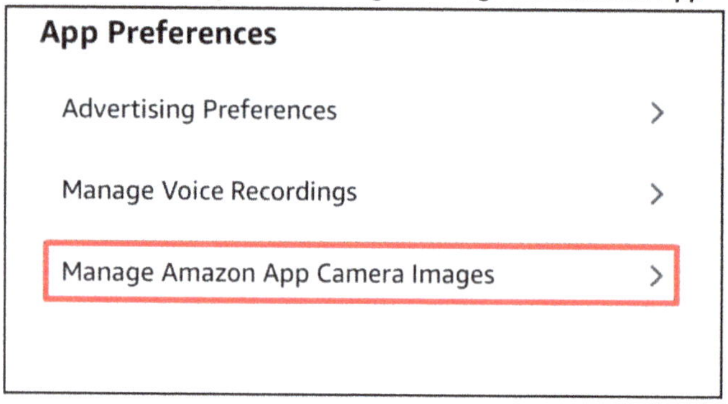

Then,

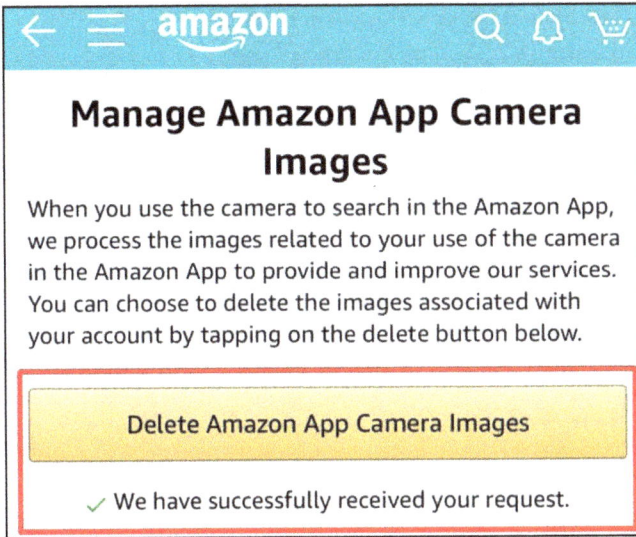

14.2 AMAZON SECURITY SETTINGS

14.2.1 Security Alerts

If you get a Security Alert about activity you do not recognize, click or tap the **Not Me** option in the notification so we can help you *reset your Amazon password* immediately to secure your account.

- If you are not able to sign into Amazon because you do not have access to the email or mobile phone on your account anymore, contact Customer Service for help restoring access.[215]

14.2.2 Browser Extensions and Privacy

Some browser extensions track your private shopping behavior and collect data like order history and items saved in your Amazon cart.

- To protect your privacy and security, please refer to the links listed below and follow the instructions supported the specific browser of your choice to remove a harmful extension.[216]

- *Chrome*
- *Firefox*
- *Safari*
- *Edge*
- *Opera*

14.2.3 Two-Step Verification

It is highly recommended you enable this feature in Amazon.

- When you try to log in, Two-Step Verification sends you a unique security code.
- Per Amazon, when you sign up for *Two-Step Verification*, Amazon will *send you a unique code by text message, voice call, or authenticator app*.
- The following *link* takes the mystery out of enrolling in this feature.

14.2.4 One-Time Passwords for All Devices

After enrolling in Two-Step Verification, I would also recommend not suppressing any future One-Time Password (OTP) challenges as this moves you from the realm of Two-Factor Authentication to a *Multi-Factor Authentication* posture within Amazon.

- This feature allows you to enable a requirement for OTP on all devices. I would highly recommend enabling this feature.

14.2.5 Secure Delivery with One-Time Password (OTP)

If you want to take your Operational Security to the next level might, I would recommend enabling *Amazon's One-time password (OTP) verification feature*.

- By enabling OTP verification, Amazon will send you a six-digit numeric PIN code that is valid until the end of the day adding yet another layer of security to your packages.
- Should you be delayed and miss the designated rendezvous point and time of package delivery, Amazon has you covered in case as they will re-attempt delivery the next day or if you have a trusted contact, you may share the OTP with whoever you choose to receive the package on your behalf.
- Remember to never share the OTP with the delivery agent over phone as OTP is intended for you to ensure secured delivery of the package.

14.2.6 1-Click Settings

1-Click lets you associate a credit, debit, or Amazon Store Card with addresses you ship to often so you can place orders with a single click of a button.

- When you disable *1-Click, it only disables 1-Click for orders that can be shipped.* 1-Click ordering does not affect digital purchases.
- Since your browser must be cookie-enabled to use 1-Click shopping, if your browser is not cookie-enabled, you can still buy items by adding them to your Shopping Cart and clicking *"Proceed to checkout"*.
- It is recommended using the *"Disable 1-Click everywhere"* setting, which you can also enable for your Mobile orders at the following *link* to ensure you do not fall victim to a scam.

14.3 AMAZON ALEXA ECHO SETTINGS

14.3.1 Delete Voice Recordings

You can play back all of the recordings in the history menu on *Alexa.amazon.com*, and if you like you can delete the recordings one-by-one.[217]

- However, if you want to remove all the recordings, the best way to do this is to visit the *"Manage your content and devices" page on Amazon.com*.
- Any Alexa apps you have registered, as well as the Echo smart speakers, will be listed on this page.
- You can select each one, and remove the recordings associated with the app or device.
- If you use Alexa on a Fire tablet, you can also remove those recordings from this page.

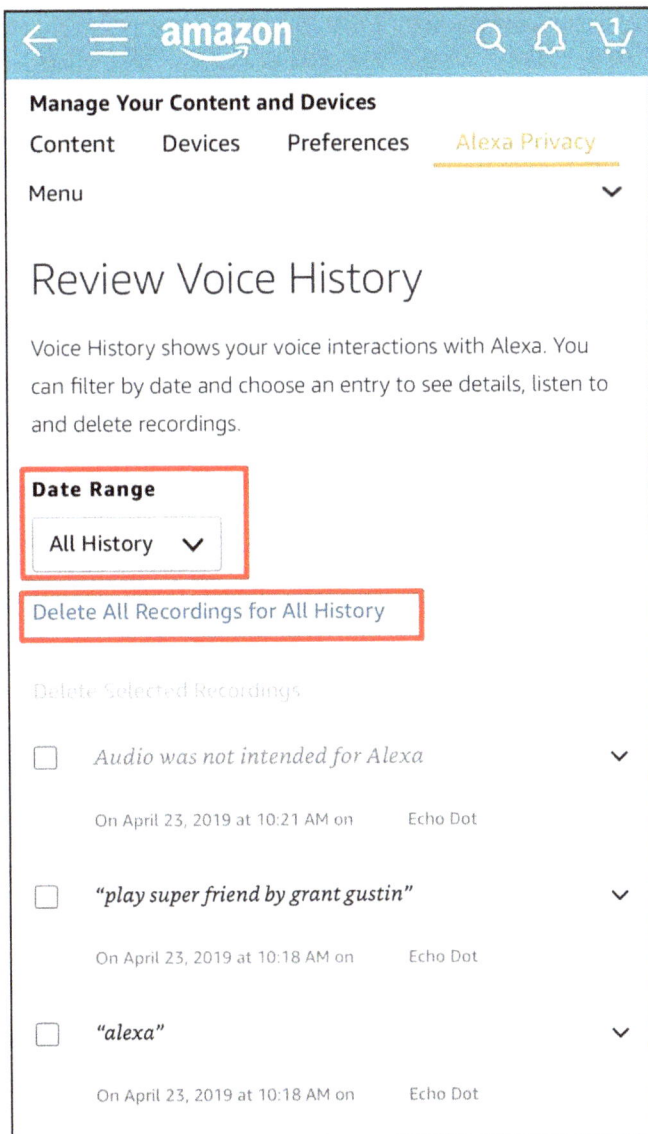

14.3.2 Disable Voice Purchasing

Go to Settings -> Alexa Account -> Voice Purchasing and make sure the **"Purchase by voice"** option is turned off to stop voice command purchases.

- If you want to use voice purchasing but want to keep others from using it, you can generate a voice code, which will be needed before every purchase.
- Alternatively, follow the instructions at URL Follow the instructions from URL *https://www.azcentral.com/story/money/business/tech/2019/04/29/heres-how-you-use-amazons-echo-google-home-apples-homepod-without-giving-up-your-privacy/3589325002/* to stop voice command purchases.

14.3.3 Managing How Your Data Improves Alexa and Opting Out

Manage How Your Data Improves Alexa

Use Voice Recordings to Help Develop New Features

Training Alexa with recordings from a diverse range of customers helps ensure Alexa works well for everyone. When this setting is enabled, your voice recordings may be used in the development of new features.

If you turn this setting off, new features may not work well for you

Help Develop New Features

Learn more about Alexa and Privacy

Use Messages to Improve Transcriptions

Allow Amazon to use messages you send with Alexa to improve transcription accuracy.

14.4 AMAZON SIDEWALK

Amazon Sidewalk is a new feature rolling out to Amazon-branded gadgets in the final weeks of 2020.
This service is designed to act as a backup network in the event Ring and Echo devices lose their internet connection.[218]

- Amazon Sidewalk allows select Echo and Ring devices to piggyback off nearby Amazon gadgets' connections.[219]
- This can include devices belonging to other people in other houses.
- If a user has *Amazon Sidewalk enabled on their Ring or Echo*, their devices can use your connections in an outage as well.
- If you have not done so already, I would highly recommend reading the *Amazon Sidewalk Privacy and Security Whitepaper*.
- While it is possible that this feature dovetails with Amazon's *"Saved Wi-Fi Password Feature"*, it is not specifically listed in the white paper and at the time of this publication.
- As of this publication, it is unknown whether opting into the use of Amazon Sidewalk re-enables this feature to use or it is simply restored by opting into Amazon Sidewalk.

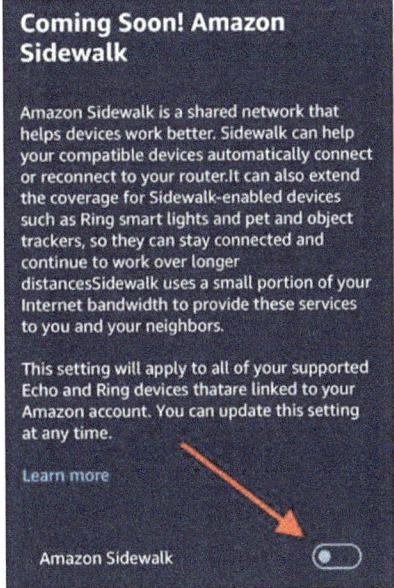

Coming Soon! Amazon Sidewalk

Amazon Sidewalk is a shared network that helps devices work better. Sidewalk can help your compatible devices automatically connect or reconnect to your router.It can also extend the coverage for Sidewalk-enabled devices such as Ring smart lights and pet and object trackers, so they can stay connected and continue to work over longer distancesSidewalk uses a small portion of your Internet bandwidth to provide these services to you and your neighbors.

This setting will apply to all of your supported Echo and Ring devices thatare linked to your Amazon account. You can update this setting at any time.

Learn more

Amazon Sidewalk

15 GAMING CONSOLES

Gaming consoles like the Nintendo Switch, PlayStation 4, and X-Box One all have social media services. Check the below settings and advice for controlling your accounts' privacy.

15.1.1 Consoles and Online Services

Service	Privacy Settings/Advice

Nintendo Switch	https://en-americas-support.nintendo.com/app/answers/detail/a_id/15987/~/how-to-adjust-nintendo-account-profile-settings-%28country%2C-email%2C-etc.%29
PlayStation 4 (PS4) and PlayStation Network (PSN)	https://www.playstation.com/en-gb/get-help/help-library/my-account/parental-controls/how-to-use-playstation-4-to-limit-who-can-contact-you-over-plays/ https://www.playstation.com/en-us/account-security/2-step-verification/ https://thenextweb.com/basics/2019/01/31/playstation-4-privacy-settings-hiding/
X-Box One (XONE) and X-Box Live	https://support.microsoft.com/en-us/help/4482922/xbox-one-online-safety-and-privacy-settings-for-parents-and-kids https://www.thewindowsclub.com/how-to-setup-xbox-privacy-and-online-safety-for-kids

16 MONEY SERVICES

Money services are unique in that their primary purpose is financial, but they also share attributes with social media, such as the ability to network and/or search for user profiles. Because the main service is financial, platform reviews and recommendations can tend to focus on security of finances, rather than privacy of personal information - but when a platform stores photos, "friends," comment history, home addresses, contact information, and more, you should protect your money service account the same way you would protect any of your other social media.

- A money service business is a non-bank institution that provides mechanisms for people to pay in any way or obtain money or cash in exchange for payment through a financial institution or institution.[220]
- An MSB provides a significant financial service to underdeveloped regions, often with limited or no banking services such as a small organization with outlets such as markets, pharmacies, and retailers.[221]
- In the United States and many other countries throughout the globe, regulations around money transmission are serious business as transmitting money is a serious business.

16.1 MONEY SERVICES SECURITY AND PRIVACY CONTROLS

When it comes to Money Services that are available to use, there are an ever-growing plethora of choices that offer unique ways to keep your money moving. The following links below provide you with the security and privacy settings a user can configure to reduce their Digital Exhaust.

Platform	Privacy Settings Link
PayPal Security	https://www.paypal.com/us/webapps/mpp/paypal-safety-and-security
PayPal Privacy	https://www.paypal.com/myaccount/privacy
Venmo Security and Privacy	https://venmo.com/account/settings/profile
CashApp Security	https://cash.app/help/us/en-us/1015-account-settings
CashApp Privacy	https://cash.app/legal/us/en-us/privacy
Braintree Security	https://www.braintreepayments.com/faq
Braintree Privacy	https://braintree.com/docs/privacy_policy.html
Google Pay Security	https://safety.google/intl/en_us/pay/
Google Pay Privacy	https://payments.google.com/legaldocument?family=0.privacynotice&hl=en-GB
Apple Pay Security	https://support.apple.com/en-us/HT203027
Apple Pay Privacy	https://support.apple.com/en-us/HT210665
Amazon Pay Security	https://paymentservices.amazon.com/docs/EN/51.html
Amazon Pay Privacy	https://paymentservices.amazon.com/privacy
Masterpass Security	https://masterpass.com/en-jp/faqs/manage-account-security.html
Masterpass Privacy	https://wallet.masterpass.com/Wallet/masterpass/en-au/privacy.html

16.2 Additional PayPal Privacy Settings

PayPal's account data and privacy settings allow users to manage the use of PayPal to make payments on other apps and websites. Within the data and privacy settings, users can also turn off various cookies and control settings such as reminders and advertisements.[222]

16.2.1 Setting Payments to Private

By default, any time you pay for something through Venmo, that amount, and description are public and shown to your other friends on the app. Here is how to make it private.

- In the smartphone app, click on the profile icon, then the settings icon (*looks like a gear*). Select **Privacy** and set the Default Privacy Settings to **Private** (not **Public** or **Friends**).

16.2.2 Hide Past Transactions

You will have made an added privacy tweak to hide your past Venmo payments.

- In the same screen, scroll down to **More** and click **Past Transactions**. Tap on **"Change All to Private"**

16.3 "Tipping" on Twitter

In May 2021, Twitter integrated a PayPal "Tip Jar" system into Twitter's website, only to receive concerns from users when it was found that Tip Jar revealed the sender's address during each transaction. [223]

- This meant that any Twitter user who "tipped" another user could unknowingly reveal where they live.
- Fortunately, this risk can be mitigated by users selecting **"No Address Needed"** as an option when they send someone a "tip" on Twitter.[224]

16.4 Additional Venmo Privacy Settings

Venmo, which is owned by PayPal[225], offers privacy settings for your transaction history as well as your user account, but it should be noted that most information is set to **"public"** by default.[226]

- Also of note, any user information sent to Venmo is accessible to PayPal as well.[227]

16.4.1 Venmo Transaction Settings

- <u>Public:</u> The transaction will be shared on the public feed and anyone on the internet may be able to see it.
- <u>Friends only:</u> The transaction will only be shared with your Venmo friends and with the other participant's Venmo friends.
- <u>Private:</u> Venmo will not share the transaction anywhere other than the "Your Stories" tab in the personal transactions feed and, if it is a payment to another user, the feed of the other person in the payment.

16.4.2 Sender/Recipient Payment Information

The payment amount, payment note, names of sender/recipient, and timestamp of the payment are available to everyone involved in the payment.

- ONLY the sender of the payment has access to the payment method used (for example: the bank account, debit/credit card number, etc.). The recipient will NEVER see this information.

16.4.3 Visibility of Payment Information

When a payment is shared, the payment notes, names of sender/recipient, and timestamp of the payment will be visible on the public feed.

- ONLY the sender and recipient have access to the payment amount.
- ONLY the sender of the payment has access to the payment method used.

16.4.4 Sharing Payments

You can set the privacy setting on a payment or purchase on an individual basis. If you do not want to change the privacy setting every time you make a payment, you can change your default privacy setting. Your future payments will automatically default to your preference, but you can adjust this before completing the payment. See instructions below on how to change your privacy setting.

- When you transact with someone else on Venmo, including payouts from merchants or payments with business profiles, the more restrictive privacy setting between the two of you will be honored. If you have your payments set to Private but your friend has their payments set to Public, a payment between the two of you will be set to Private.
- Purchases made using your Venmo MasterCard Debit Card or Venmo Credit Card, and purchases from approved merchants when you pay with Venmo are Private by default, but you can change the privacy setting on any purchase to share them.
- All your transactions, regardless of privacy setting, will still be visible in your personal transactions feed so that you have a transaction record.

16.4.5 Privacy Settings Individual Payment or Purchase

You can set the privacy setting for each individual payment or purchase, right from the payment or buy itself.

- Just select or tap on the privacy setting in any payment or purchase and select your preferred setting.
- Venmo's privacy webpage explains that transactions where each party has different settings, the more restricted setting will always be used[228]—so ensure you are protected by changing your default privacy setting to "private".

16.4.6 Hiding Past and Future Transactions

If you have not been setting individual transactions to **"private"** as you go, you can still hide your entire history with a few clicks.

- First, navigate to your home page, and then select **"Settings"** from the sidebar.

- From Settings, select **"Privacy"**.
- Once on your Privacy Settings page, set your Default Privacy Settings to **"Private."** To hide your entire transaction history, select **"Change All to Private"** in the **"Past Payments"** section.

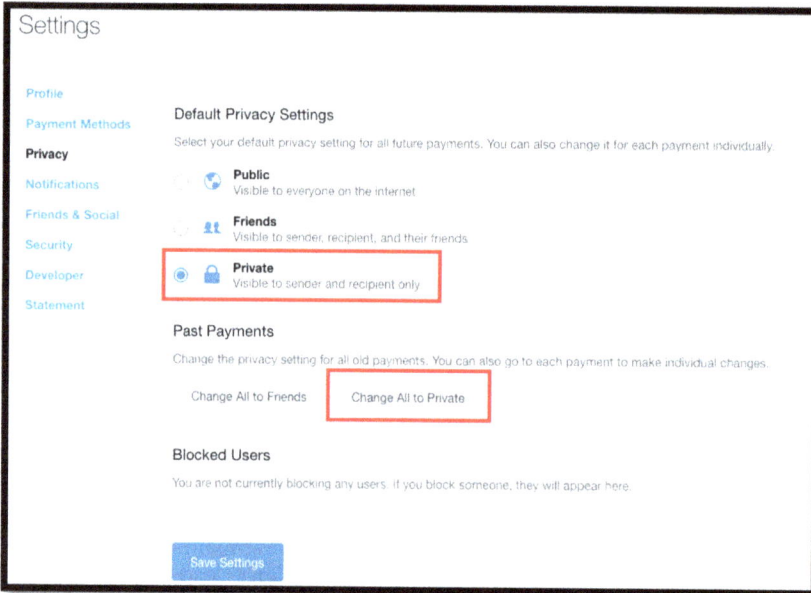

17 PHOTO METADATA

Photo metadata are set of data describing and supplying information about rights and administration of an image.

- Many devices with cameras, like smartphones, embed the set of data into the pictures they capture.
- Data types include the shutter speed, ISO, aperture data, camera mode, and/or GPS location of where the picture was taken.
- They are stored within the pictures they take in a format called the Exchangeable Image Format (EXIF) and left intact, present a potential privacy vulnerability when shared across devices or uploaded onto the Internet. In short, to protect your privacy, remove EXIF data from your images.

17.1 IOS

17.1.1 Remove EXIF data
Prior to Apple's release of iOS 13 there was no native way to disable EXIF data.

- With the release of iOS 14, Apple now supplies users a way to remove EXIF data from photos. This **URL** will inform you on how to do so along with other key features within iOS 15 that will better enhance your privacy.
- However, apps, which can remove EXIF data, are available in the iOS App Store.
- One such app includes Exif Data and the pro version costs $0.99/year. It enables you to view, edit, and remove metadata from your iOS devices like iPhone and iPad.
- It also allows you to spoof a location of your choosing of where the photo was taken which will appear within the photo's metadata.

Figure 7. Icon for Exif Metadata Apo

17.1.2 EXIF iOS photos on Apple Mac

According to the article *"Parenting tip: Share your iOS photos without revealing your EXIF location data"* at URL https://www.engadget.com/amp/2013/10/03/parenting-tip-share-your-ios-photos-without-revealing-your-exif/ :"

- The easiest way to view EXIF data is on your Mac. Just transfer your photos to your Mac using **iPhoto**, tap on the image and select the "**i**" for info.
- All the EXIF data, including a map of the GPS coordinates will appear within the iPhoto window.
- If you do not see a map, then you may have to hop into **iPhoto preferences** and **turn on** this mapping feature. **Go to iPhoto > Preferences, and then click Advanced**. If you choose "**Automatically**", then iPhoto will scan your photos for GPS data and map them for you.
- While you are in the settings, you should check the status of the "**Include location information for published photos**" option.
- If it is selected, then the location data will remain intact when you use iPhoto to upload your photos to other services.
- If it is not selected, then the location data will be stripped from the file by iPhoto during the upload process.

- Unselecting this option is the preferred choice if you don't want people to know the location of

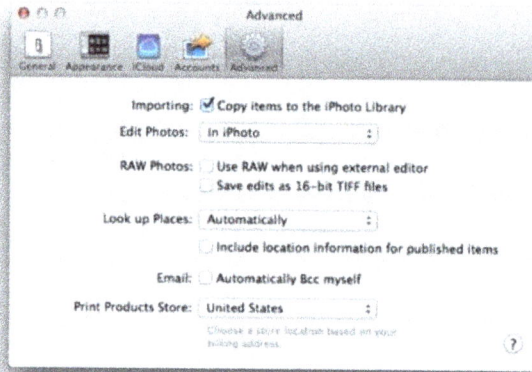

your photos."

17.1.3 EXIF Location Data on iOS

Turn off photo geotagging feature by going to **Location Services** in the **Settings**.

- Tap on **Settings > Privacy > Location Services** and then scroll down to the **Camera app** to make sure it is toggled off.
- NOTE: this only applies to photos taken after you have turned off the location feature and does not remove any other EXIF data.

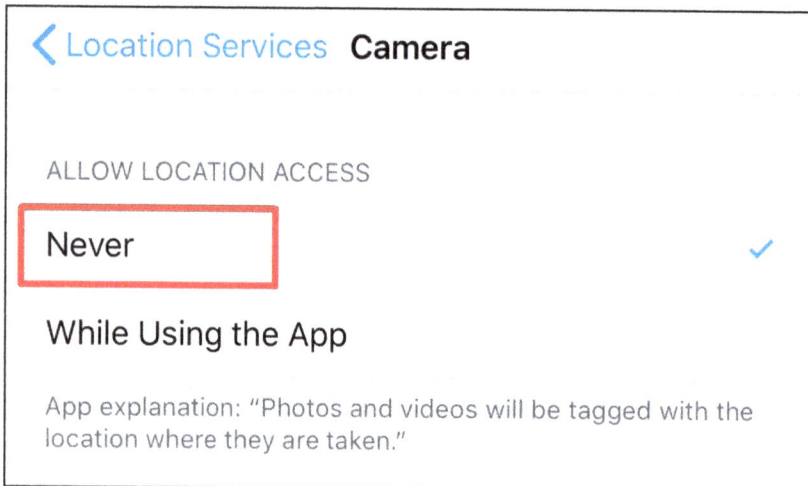

17.1.4 iOS App Change Camera Settings

Enable Screen Time for your devices, go to Location Services, and click Don't Allow changes.

- See Section 3.18.1 for more information about Screen Time.
- You can also visit Apple's information about Screen Time at *URL* *https://support.apple.com/en-us/HT208982*

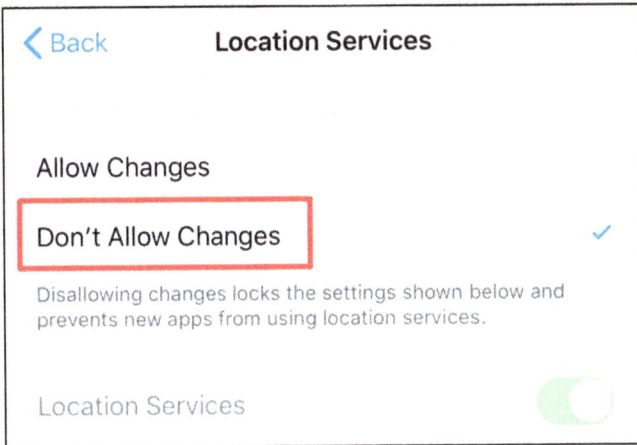

17.2 ANDROID

17.2.1 Camera App Location Data

Open the **Camera app** on your phone.

- Tap the **Settings** option on the viewfinder. For Samsung phones, the settings gear is in the top left corner. For Google Pixel phones, you will need to tap the downward-facing arrow at the top of the screen, and then tap the settings gear in the menu that appears.
- Turn off the **Location toggle** in the setting menu. On Samsung phones, Location is near the bottom, but it is the first setting in Google Camera advanced menu.[229]

17.2.2 Gallery App Location Data

Open **Gallery app** on your phone.

- Tap the **picture** you want to remove location data from.
- **Swipe up** on the picture to pull up the picture's information.
- Tap **Edit**.
- Tap the **red minus** next to the location data to remove it.
- Tap **Save**.

17.3 GOOGLE PHOTOS

There is an obvious concern any time you upload your pictures to a service on the internet you should exercise caution.[230] Even though Google actively works to secure their services, there is always a chance of vulnerability and the risk that someone could get access to your pictures and videos. The following privacy settings are worth noting should you choose to enable them:

- Only share pictures with people you know.
- Check the **"Sharing"** settings on each album you create.
- Do not upload pictures to Shared Albums from people you do not know.
- Turn on **"Remove Geo-Location in Items Shared by Link"**.
- Turn off "**Google Location History**" in the Google Photos Settings.
- Occasionally check the Sharing settings on your account to keep things private.

Beyond what was noted above, Google has other specific privacy settings available with Google Photos.

17.3.1 Location Data in Photos

Open **Google Photos** on your phone or visit the *Google Photos website* on your computer.

- Open the **picture** you wish to remove location data from.
- In the Google Photos app, **swipe up** to reveal the photo information. On desktop, click the **Info** icon in the top right option bar (looks like a lower case *i* in a circle).
- Tap the **icon** to the right of the listed location.
- In the Google Photos app, tap **Remove Location**. On desktop, click **No location**.
- In the Google Photos app, tap **Remove**.

17.3.2 Memories

Memories are collections of some of your best photos and videos whether from previous years or recent weeks. Memories are available on Android devices, iPhones, and iPads.

You can select the types of Memories you want to see above your photo grid. The Memories carousel above the photo grid only appears when at least one memory type is selected.[231]

- On your Android phone or tablet, open the Photos app Photos.
- At the top right, tap your account profile photo or initial and then Photo's settings and then Memories.
- Tap Featured memories.
- Select the types of memories you want to see.

17.3.3 Hide someone

Google allows you to exclude people and even pets from Memories.[232]

- On your Android phone or tablet, open the Photos app Photos.
- At the top right, tap your account profile photo or initial and then Photo's settings and then Memories.
- Tap Hide people & pets.
- Choose who you want to hide.
- To show someone, tap their face again.

18 TRAFFIC LIGHT PROTOCOL (TLP) DEFINITIONS

Traffic Light Protocol (TLP) Definitions

Color	When should it be used?	How may it be shared?
TLP:RED t for disclosure, restricted to participants only.	Sources may use TLP:RED when information cannot be effectively acted upon by additional parties, and could lead to impacts on a party's privacy, reputation, or operations if misused.	Recipients may not share TLP:RED information with any parties outsid of the specific exchange, meeting, or conversation in which it was originally disclosed. In the context of a meeting, for example, TLP:REI information is limited to those present at the meeting. In most circumstances, TLP:RED should be exchanged verbally or in person.
TLP:AMBER nited disclosure, restricted to participants' organizations.	Sources may use TLP:AMBER when information requires support to be effectively acted upon, yet carries risks to privacy, reputation, or operations if shared outside of the organizations involved.	Recipients may only share TLP:AMBER information with members of their own organization, and with clients or customers who need to know the information to protect themselves or prevent further harm. **Sources are at liberty to specify additional intended limits of the sharing: these must be adhered to.**
TLP:GREEN nited disclosure, restricted to the community.	Sources may use TLP:GREEN when information is useful for the awareness of all participating organizations as well as with peers within the broader community or sector.	Recipients may share TLP:GREEN information with peers and partner organizations within their sector or community, but not via publicly accessible channels. Information in this category can be circulated wide within a particular community. TLP:GREEN information may not be released outside of the community.
TLP:WHITE)isclosure is not limited.	Sources may use TLP:WHITE when information carries minimal or no foreseeable risk of misuse, in accordance with applicable rules and procedures for public release.	Subject to standard copyright rules, TLP:WHITE information may be distributed without restriction.

19 ENDNOTES

[1] Website | DigitalElement.com | "Digital Data Exhaust – 2018 Research Results" | 2021 | https://www.digitalelement.com/digital-data-exhaust/ | accessed on 20 June 2021.

[2] Online article | Norton | "How Data Brokers Find and Sell Your Personal Info" | 18 January 2021 | https://us.norton.com/internetsecurity-privacy-how-data-brokers-find-and-sell-your-personal-info.html | accessed on 22 July 2021.

[3] Online article | TechCrunch | "The Power of Data Exhaust" | 26 May 2013 | https://techcrunch.com/2013/05/26/the-power-of-data-exhaust/ | accessed on 20 June 2021.

[4] Online article | CultureBy – Grant McCracken | "How Social Networks Work: The Puzzle of Exhaust Data" | 19 July 2007 | https://cultureby.com/2007/07/how-social-netw.html | accessed on 20 June 2021.

[5] Online article | Ghostery | "Defining Your Digital Footprint" | 3 March 2020 | https://www.ghostery.com/defining-your-digital-footprint/ | accessed on 20 June 2021.

[6] Online article | Medium | "Your Data is a Myth" | 11 June 2020 | https://medium.com/swlh/your-data-is-a-myth-37997abcc82a | accessed on 20 June 2021.

[7] Online article | TNW | "Here's how we take back control over our digital identities" | 24 November 2018 | https://thenextweb.com/news/heres-how-we-take-back-control-over-our-digital-identities | accessed on 27 July 2021.

[8] Online article | Ghostery | "The Privacy Paradox: Concerns vs Behaviors" | 24 March 2020 | https://www.ghostery.com/the-privacy-paradox-concerns-vs-behavior/ | accessed on 20 June 2021.

[9] Online article | Privacy News Online | "When the Home is no Data Protection Haven: Addressing Privacy Threats from Intimate Relationships" | 12 June 2020 | https://www.privateinternetaccess.com/blog/?p=13298 | accessed on 20 June 2021.

[10] Online article | Ghostery | "Defining Your Digital Footprint" | 3 March 2020 | https://www.ghostery.com/defining-your-digital-footprint/ | accessed on 27 July 2021.

[11] Online article | Ghostery | "Digital Self Infographic" | 31 March 2020 | https://www.ghostery.com/digital-self-summary-infographic/ | accessed on 26 July 2021.

[12] Online article | Ghostery | "Simple Steps to Manage Your Digital Footprint" | https://www.ghostery.com/easy-ways-to-manage-your-digital-footprint/ | accessed on 27 July 2021.

[13] Online article | Visual Capitalist | "The Multi-Billion Dollar Industry That Makes Its Living from Your Data" | 14 April 2018 | https://www.visualcapitalist.com/personal-data-ecosystem/ | accessed on 22 June 2021.

[14] Online article | Tech Target | "The Wide Web of Nation-State Hackers Attacking the US" | 20 April 2021 | https://searchsecurity.techtarget.com/news/252499613/The-wide-web-of-nation-state-hackers-attacking-the-US | accessed on 30 July 2021.

[15] Online article | Medium | "Targeted Threat Intelligence, generated from Open-Source Information" | 1 June 2020 | https://alex-newman.medium.com/targeted-threat-intelligence-generated-from-open-source-information-e13a3a2c58dd | accessed on 30 July 2021.

[16] Online article | Norton | "The Privacy Paradox: How Much Privacy Are We Willing to Give Up Online?" | 14 April 2021 | https://us.norton.com/internetsecurity-privacy-how-much-privacy-we-give-up.html | accessed on 2 August 2021.

[17] Blog post | Data Privacy | "What is Personally Identifiable Information (PII) and What is Personal Data?" | 23 February 2021 | https://dataprivacymanager.net/what-is-personally-identifiable-information-pii/ | accessed on 2 August 2021.

[18] Online article | TechCrunch | "Data Was The New Oil, Until The Oil Caught Fire" | 2 May 2021 | http://feedproxy.google.com/~r/Techcrunch/~3/1uhKSJT1A3s/ | accessed on 2 August 2021.

[19] Online article | Digital Journal | "Digital Identity and Privacy Issues are Worrying Consumers More Than Ever" | 15 May 2021 | https://www.digitaljournal.com/tech-science/digital-identity-and-privacy-issues-are-worrying-consumers-more-than-ever/article | accessed on 2 August 2021.

[20] Online article | Gizmodo | "How to Track the Tech That's Tracking You Every Day" | 6 June 2020 | https://gizmodo.com/how-to-track-the-tech-thats-tracking-you-every-day-1843908029 | accessed on 2 August 2021.

[21] Online article | MySudo | "What is Digital Exhaust and Why Does it Matter?" | 17 August 2020 | https://mysudo.com/2020/08/what-is-digital-exhaust-and-why-does-it-matter/ | accessed on 2 August 2021.

²² Blog post | Epam | "The Privacy Web: A Look into Where Personal Data Protection Stands Today and Where It's Headed" | 7 October 2020 | https://www.epam.com/insights/blogs/the-privacy-web-personal-data-protection-today-and-where-its-headed | accessed on 2 August 2021.

²³ Online article | Forbes | "Four Steps to Take Control of Your Privacy Management" | 30 March 2020. https://www.forbes.com/sites/forbestechcouncil/2020/03/30/four-steps-to-take-control-of-your-privacy-management/?sh=2fd953004761

²⁴ Online article | The Register | "Bad News: So Much of Your Personal Data has Been Hacked that Lesson Manuals on How To Use It are the Latest Hot Property" | 16 April 2020 | https://www.theregister.com/2020/04/16/cybercrimeby_fraud_lessons/ | accessed on 23 July 2021.

²⁵ Online article | AddExchanger | "Oracle Signs On To Support Unified ID 2.0" | 13 May 2021 | https://www.adexchanger.com/online-advertising/oracle-signs-on-to-support-unified-id-2-0/?utm_source=feedburner&utm_medium=feed&utm_campaign=Feed%3A+ad-exchange-news+%28AdExchanger.com%3A+Exchanging+Ideas+On+Digital+Media+Optimization%29 | accessed 22 July 2021.

²⁶ Online article | TechCrunch | "Oracle's BlueKai Tracks You Across the Web. That Data Spilled Online" | 19 June 2020 | https://techcrunch.com/2020/06/19/oracle-bluekai-web-tracking/ | accessed on 22 July 2021.

²⁷ Online article | Digiday | "Digiday Guide: Everything You Need to Know About the End of the Third-Party Cookie" | 6 May 2020 | https://digiday.com/media/digiday-guide-everything-you-need-to-know-about-the-end-of-the-third-party-cookie/ | accessed on 30 July 2021.

²⁸ Online article | PrivacyBee | https://privacybee.com/blog/what-is-browser-fingerprinting-and-how-do-i-prevent-it/ | accessed on 4 October 2021

²⁹ Online article | Privacy News Online | "Mozilla Study Reaffirms that Internet History Can be Used for "Reidentification"" | 31 August 2020 |https://www.privateinternetaccess.com/blog/mozilla-study-reaffirms-that-internet-history-can-be-used-for-reidentification/ | accessed on 25 July 2021.

³⁰ Online article | HackRead | "Cross-browser Tracking Vulnerability Compromises User Anonymity" | 17 May 2021 | https://www.hackread.com/cross-browser-tracking-compromises-user-anonymity/ | accessed on 25 July 2021.

³¹ Online article | TechCrunch | "Who Gets to Own Your Digital Identity?" | 22 August 2019 | https://techcrunch.com/2019/08/22/who-gets-to-own-your-digital-identity/ | accessed on 27 July 2021.

³² Online article | MacRumors | "A Day in the Life of Your Data: Apple Details How Companies Can Track You Across Apps and Websites" | 27 January 2021 | https://www.macrumors.com/2021/01/28/apple-a-day-in-the-life-of-your-data/ | accessed on 30 July 2021.

³³ Online article | Digiday | "Question of the Day: How are Publishers Using Contextual Data?" | 2021 | https://digiday.com/sponsored/question-of-the-day-how-are-publishers-using-contextual-data/ | accessed on 30 July 2021.

³⁴ E-book | Federal Trade Commission | "Data Brokers: A Call for Transparency and Accountability: A Report of the Federal Trade Commission" | May 2014 | https://www.ftc.gov/reports/data-brokers-call-transparency-accountability-report-federal-trade-commission-may-2014 | accessed on 5 July 2021.

³⁵ Website | Vice.com | "Here's a Long List of Data Broker Sites and How to Opt-Out of Them" | 27 March 2018 | https://www.vice.com/en_us/article/ne9b3z/how-to-get-off-data-broker-and-people-search-sites-pipl-spokeo | accessed on 28 June 2021.

³⁶ Online article | Motherboard | "What Are 'Data Brokers,' and Why Are They Scooping Up Information About You?" | 27 March 2018 | https://www.vice.com/en_us/article/bjpx3w/what-are-data-brokers-and-how-to-stop-my-private-data-collection | accessed on 28 June 2021.

³⁷ Online article | Fast Company | "Here are the Data Brokers Quietly Buying and Selling your Personal Information" | 2 March 2019 | https://www.fastcompany.com/90310803/here-are-the-data-brokers-quietly-buying-and-selling-your-personal-information | accessed on 28 June 2021.

³⁸ Online article | ProPublica | "How to Wrestle Your Data from Data Brokers, Silicon Valley — and Cambridge Analytica" | 30 April 2018 | https://www.propublica.org/article/how-to-wrestle-your-data-from-data-brokers-silicon-valley-and-cambridge-analytica | accessed on 28 June 2021.

³⁹ Online article | Quartz | "The Nine Companies That Know More About You Than Google or Facebook" | 27 May 2014 | https://qz.com/213900/the-nine-companies-that-know-more-about-you-than-google-or-facebook/ | accessed on 28 June 2021.

⁴⁰ Online article | Donald W. Reynolds National Center for Business Journalism | "The Business of Personal Data Brokers and Online Privacy" | 2 August 2018 | https://businessjournalism.org/2018/08/the-business-of-personal-data-brokers-and-online-privacy/ | accessed on 28 June 2021.

⁴¹ Website | Transparencymarketresearch.com | "Data Broker Market" | https://www.transparencymarketresearch.com/data-brokers-market.html | accessed on 5 July 2021.

[42] E-Book | Federal Trade Commission | "Data Brokers: A Call for Transparency and Accountability: A Report of the Federal Trade Commission" | May 2014 | https://www.ftc.gov/reports/data-brokers-call-transparency-accountability-report-federal-trade-commission-may-2014 | accessed on 5 July 2021.

[43] Website | E-verify.com | "About E-Verify" | 10 April 2018 | https://www.e-verify.gov/about-e-verify | accessed on 5 July 2021.

[44] Website | E-verify.com | "Self-Lock" | 7 May 2020 | https://www.e-verify.gov/mye-verify/self-lock | accessed on 5 July 2021.

[45] Online article | Mashable | "How to Keep Yourself Safe in a World of Creepy Websites Filled with Personal Data" | 11 January 2017 | https://mashable.com/2017/01/11/online-security-family-tree-now/ | accessed on 30 June 2021.

[46] Online article | People | "Here's Why This Website That Lets Anyone Find Your Address, Phone Number and More Is Scaring People" | 12 January 2017 | https://people.com/tech/this-website-lets-anyone-find-your-address-phone-number-and-more-for-free-heres-how-to-opt-out/ | accessed on 30 June 2021.

[47] Online article | Gizmodo | "When a Stranger Decides to Destroy Your Life" | 26 July 2018 | https://gizmodo.com/when-a-stranger-decides-to-destroy-your-life-1827546385 | accessed on 30 June 2021.

[48] Online article | Lifehacker | "How to Opt Out of the Most Popular People Search Sites" | 4 March 2021 | https://lifehacker.com/how-to-opt-out-of-the-most-popular-people-search-sites-1791536533 | accessed on 30 June 2021.

[49] Online article | EFF | "Protecting Your Privacy if Your Phone is Taken Away" | 4 June 2020 | https://www.eff.org/deeplinks/2020/06/protecting-your-privacy-if-your-phone-taken-away | accessed on 20 July 2021.

[50] Online article | HackRead | "How to Protect Your Privacy on a Smartphone: 12 Tips & Tricks" | 17 June 2021 | https://www.hackread.com/how-to-protect-privacy-smartphone-12-tips-tricks/ | accessed on 20 July 2021.

[51] Online article | Prey | "Phone Security: 20 Ways to Secure Your Mobile Phone" | 6 April 2021 | https://preyproject.com/blog/en/phone-security-20-ways-to-secure-your-mobile-phone/ | accessed on 20 July 2021.

[52] Online article | Wired | "Mobile Websites Can Tap into Your Phone's Sensors Without Asking" | 26 September 2018 | https://www.wired.com/story/mobile-websites-can-tap-into-your-phones-sensors-without-asking | accessed on 30 June 2021.

[53] Online article | Gizmodo | "All the Sensors in Your Smartphone, and How They Work" | 29 June 2020 | https://gizmodo.com/all-the-sensors-in-your-smartphone-and-how-they-work-1797121002 | accessed on 30 June 2021.

[54] Online article | Lifehacker | "It's Time to Check Which Apps Are Tracking Your Location" | 10 December 2018 | https://lifehacker.com/its-time-to-check-which-apps-are-tracking-your-location-1830979707 | accessed on 30 June 2021.

[55] Journal article | *Science News*, Volume 193, Number 2 | "The Spy in Your Pocket" | 3 February 2018 | https://www.sciencenews.org/sn-magazine/february-3-2018 | accessed on 5 July 2021.

[56] Online article | Digital Trends | "How to Track a Phone Using Android or iOS" | 21 March 2021 | https://www.digitaltrends.com/mobile/how-to-track-a-cell-phone/ | accessed on 30 June 2021.

[57] Online article | Lifehacker | "It's Time to Check Which Apps Are Tracking Your Location" | 10 December 2018 | https://lifehacker.com/its-time-to-check-which-apps-are-tracking-your-location-1830979707 | accessed on 30 June 2021.

[58] Blog post | AppCensus | "Ad IDs Behaving Badly" | 14 February 2019 | https://blog.appcensus.mobi/2019/02/14/ad-ids-behaving-badly | accessed 30 June 2021.

[59] Online article | Lifehacker | "PSA: Your Phone Logs Everywhere You Go. Here's How to Turn It Off" | 9 August 2019 | https://lifehacker.com/psa-your-phone-logs-everywhere-you-go-heres-how-to-t-1486085759 | accessed on 30 June 2021.

[60] Online article | Gizmodo | "How Location Tracking Actually Works on Your Smartphone" | 3 September 2018 | https://gizmodo.com/how-location-tracking-actually-works-on-your-smartphone-1828356441 | accessed on 30 June 2021.

[61] Blog post | BuzzFeed | "A Lot of Apps Sell Your Data. Here's What You Can Do About It" | 1 May 2018 | https://www.buzzfeednews.com/article/nicolenguyen/how-apps-take-your-data-and-sell-it-without-you-even | accessed on 30 June 2021.

[62] Online article | Apple | "Set a passcode on iPhone" | https://support.apple.com/guide/iphone/set-a-passcode-iph14a867ae/ios | accessed on 1 October 2021

[63] Online article | Apple | "Inspecting App Activity Data" | https://developer.apple.com/documentation/foundation/urlrequest/inspecting_app_activity_data | accessed on 1 October 2021

[64] Online article | Apple | "Two-factor authentication for Apple ID" | https://support.apple.com/en-us/HT204915 | accessed on 1 October 2021

[65] Website | MacRumors | "iOS 15: How to Hide Your IP Address From Trackers in Safari" | https://www.macrumors.com/how-to/hide-your-ip-address-from-trackers-ios/ | accessed on 1 October 2021

[66] Website | FastCompany | "Use these 8 new iOS 15 privacy and security features right away" | https://www.fastcompany.com/90673312/ios-15-iphone-privacy-security-features | accessed on 1 October 2021

[67] Website | MacRumors | "Apple Putting a Stop to Email Tracking Pixels With Mail Privacy Protection in iOS 15 and macOS Monterey" | https://www.macrumors.com/2021/06/10/ios-15-mail-privacy-protection-tracking-pixels/ | accessed on 1 October 2021

68 Online article | Digiday | "The Elephant in the Room: Companies Persist with Fingerprinting as a Workaround to Apple's New Privacy Rules" | 12 April 2021 | https://digiday.com/media/the-elephant-in-the-room-companies-persist-with-fingerprinting-as-a-workaround-to-apples-new-privacy-rules/ | accessed on 8 August 2021.

69 Online article | Future of Privacy Forum | "Understanding the World of Geolocation Data" | 22 May 2020 | https://fpf.org/blog/understanding-the-world-of-geolocation-data/ | accessed on 30 July 2021.

70 Online article | 9to5Mac | "iOS 14 Lets Users Grant Approximate Location Access for Apps That Don't Require Exact GPS Tracking" | 12 August 2020 | https://9to5mac.com/2020/08/12/ios-14-precise-location/ | accessed on 30 July 2021.

71 Press Release | NSA | "How Mobile Device Users Can Limit Their Location Data Exposure" | 4 August 2020 | https://www.nsa.gov/news-features/press-room/Article/2298930/how-mobile-device-users-can-limit-their-location-data-exposure/ | accessed on 30 July 2021.

72 Online article | MacRumors | "Security Researchers Develop Framework for Tracking Bluetooth Devices Using Find My" | 4 March 2021 | https://www.macrumors.com/2021/03/04/security-researchers-find-my-tracking-framework/ | accessed on 23 July 2021.

73 Online article | How-To Geek | "What Is Apple's Find My Network?" | 9 May 2021 | https://www.howtogeek.com/725842/what-is-apples-find-my-network/| accessed 27 July 2021.

74 Website | MacRumors | "iOS 15 Overview" | https://www.macrumors.com/roundup/ios-15/ | accessed on 1 October 2021

75 Online article | Android Authority | "Hands-on with Privacy Dashboard, One of Android 12's Best New Features" | 9 June 2021 | https://www.androidauthority.com/android-privacy-dashboard-1233846/ | accessed on 5 August 2021.

76 Online article | DefendingDigital | "Android Security And Privacy Guide 2021" | 2021 | https://defendingdigital.com/android-security-privacy-guide/ | accessed on 5 August 2021.

77 Online article | DefendingDigital | "Android Security And Privacy Guide 2021" | 2021 | https://defendingdigital.com/android-security-privacy-guide/ | accessed on 5 August 2021.

78 Website | Support.Google.com | "Play Console Help" | 2021 | https://support.google.com/googleplay/android-developer/answer/6048248?hl=en | accessed on 8 August 2021.

79 Online article | Android Authority | "Hands-on with Privacy Dashboard, One of Android 12's Best New Features" | 9 June 2021 | https://www.androidauthority.com/android-privacy-dashboard-1233846/ | accessed on 5 August 2021.

80 Online article | Restore Privacy | "How to Secure Your Android Device and Have More Privacy" | 13 May 2020 | https://restoreprivacy.com/secure-android-privacy/ | accessed on 5 August 2021.

81 Online article | Android Central | "How to remove location data from photos on Android" | 3 June 2020 | https://www.androidcentral.com/how-remove-location-data-photos-android | accessed on 8 August 2021.

82 Online article | Android Central | "How to remove location data from photos on Android" | 3 June 2020 | https://www.androidcentral.com/how-remove-location-data-photos-android | accessed on 8 August 2021.

83 Online article | How-to Geek | "What Is the Privacy Dashboard on Android?" | 14 June 2021 | https://www.howtogeek.com/733712/what-is-the-privacy-dashboard-on-android/ | accessed on 5 August 2021.

84 Online article | PrivacySavvy | "Disable Ad Tracking on All Your Devices: The Complete Guide with Screenshots" | 27 May 2021 | https://privacysavvy.com/security/safe-browsing/disable-ad-tracking/ | accessed on 5 August 2021.

85 Website | Support.Google.com | "Play Console Help" | 2021 | https://support.google.com/googleplay/android-developer/answer/6048248?hl=en | accessed on 8 August 2021.

86 Website | Support.Google.com | "Play Console Help" | 2021 | https://support.google.com/googleplay/android-developer/answer/6048248?hl=en | accessed on 8 August 2021.

87 Online article | Make Use Of | "How Do Websites Track Your Online Activities?" | 22 May 2021 | https://www.makeuseof.com/how-do-websites-track-your-online-activities/ | accessed on 24 July 2021.

88 Online article | "Now Sites Can Fingerprint You Online Even When You Use Multiple Browsers" | 13 February 2017 | https://arstechnica.com/information-technology/2017/02/now-sites-can-fingerprint-you-online-even-when-you-use-multiple-browsers/ | accessed 24 July 2021.

89 Online article | Fox News | "Lock Down Your Phone From Snoops and Hackers" | 12 June 2021 | https://www.foxnews.com/tech/lock-down-phone-from-snoops-hackers | accessed on 20 July 2021.

90 Online article | Vice | "Inside the Industry That Unmasks People at Scale" | 14 July 2021 | https://www-vice-com.cdn.ampproject.org/c/s/www.vice.com/amp/en/article/epnmvz/industry-unmasks-at-scale-maid-to-pii | accessed 22 July 2021.

91 Online article | Privacy News Online | "Keeping Your Digital Footprint Clean During Quarantine" | 29 April 2020 | https://www.privateinternetaccess.com/blog/keeping-your-digital-footprint-clean-during-quarantine/ | accessed on 26 July 2021.

92 Blog post | Ghostery | "Behind the Curtain" | 2021 | https://www.ghostery.com/tag/digital-self/ | accessed 26 July 2021.

93 Online article | TechCrunch | "Who Gets to Own Your Digital Identity?" | 22 August 2019 | https://techcrunch.com/2019/08/22/who-gets-to-own-your-digital-identity/ | accessed on 27 July 2021.

94 Online article | Future of Privacy Forum | "A Closer Look at Location Data: Privacy and Pandemics" | 25 March 2020 | https://fpf.org/blog/a-closer-look-at-location-data-privacy-and-pandemics/ | accessed on 30 July 2021.

95 Online article | Naked Security by Sophos | "Woman Stalked by Sandwich Server via Her COVID-19 Contact Tracing Info" | https://nakedsecurity.sophos.com/2020/05/14/woman-stalked-by-sandwich-server-via-her-covid-19-contact-tracing-info/ | accessed on 30 July 2021.

96 Press Release | NSA | "How Mobile Device Users Can Limit Their Location Data Exposure" | 4 August 2020 | https://www.nsa.gov/news-features/press-room/Article/2298930/how-mobile-device-users-can-limit-their-location-data-exposure/ | accessed on 30 July 2021.

97 Online article | Future of Privacy Forum | "New Infographic Illustrates Key Aspects of Location Data" | 22 May 2020 | https://fpf.org/press-releases/new-infographic-illustrates-key-aspects-of-location-data/ | accessed on 30 July 2021.

98 Online article | Wonder How To | "Find Identifying Information from a Phone Number Using OSINT Tools" | 7 June 2019 | https://null-byte.wonderhowto.com/how-to/find-identifying-information-from-phone-number-using-osint-tools-0195472/ | accessed on 2 July 2021.

99 Online article | Gizmodo | "Your Old Phone Number Could Get You Hacked, Researchers Say" | 3 May 2021 | https://gizmodo.com/your-old-phone-number-could-get-you-hacked-researchers-1846813781 | accessed on 2 July 2021.

100 Online article | Gizmodo | "Your Old Phone Number Could Get You Hacked, Researchers Say" | 3 May 2021 | https://gizmodo.com/your-old-phone-number-could-get-you-hacked-researchers-1846813781 | accessed on 2 August 2021.

101 Online article | Gizmodo | "Your Old Phone Number Could Get You Hacked, Researchers Say" | 3 May 2021 | https://gizmodo.com/your-old-phone-number-could-get-you-hacked-researchers-1846813781 | accessed on 30 July 2021.

102 Online article | Safeguarde | "How to Block Restricted Calls on Android and iPhone" | 2021 | https://safeguarde.com/how-to-block-restricted-calls-on-android-and-iphone/ |accessed on 20 July 2021.

103 Online article | NSA | "NSA Issues Guidance on Securing Wireless Devices in Public Settings" | 29 July 2021 | https://www.nsa.gov/News-Features/Feature-Stories/Article-View/Article/2711968/nsa-issues-guidance-on-securing-wireless-devices-in-public-settings/ | accessed on 25 August 2021.

104 Cybsersecurity Information Sheet | NSA | "Securing Wireless Devices in Public Settings" | July 2021 | https://media.defense.gov/2021/Jul/29/2002815141/-1/-1/0/CSI_SECURING_WIRELESS_DEVICES_IN_PUBLIC.PDF | accessed on 25 August 2021.

105 Online article | Norton | "The Dangers of Public Wi-Fi" | 26 May 2018 | https://us.norton.com/internetsecurity-privacy-risks-of-public-wi-fi.html | accessed on 25 August 2021.

106 Online article | AT&T Business | "WPA Security Explained: What is Wi-Fi Protected Access?" | 29 June 2020 | https://cybersecurity.att.com/blogs/security-essentials/wpa-security-explained-what-is-wi-fi-protected-access | accessed on 23 July 2021.

107 Online article | PixelPrivacy | "The Real Life Dangers of Using Public Wi-Fi (and How to Protect Yourself When You Have to Use it)" | 23 May 2021 | https://pixelprivacy.com/resources/public-wifi-dangers/ | accessed on 2 July 2021.

108 Online article | Cloudwards | "Dangers of Public WiFi: What You Need to Know in 2021" | 8 June 2021 | https://www.cloudwards.net/dangers-of-public-wifi/ | accessed on 25 August 2021.

109 Online article | Norton | "The dos and don'ts of using public Wi-Fi" | 23 July 2018 | https://us.norton.com/internetsecurity-wifi-the-dos-and-donts-of-using-public-wi-fi.html | accessed on 25 August 2021.

110 Online article | PixelPrivacy | "Recap: The Best Ways to Protect Your Public Wi-Fi Connection" | 23 July 2021 | https://pixelprivacy.com/resources/public-wifi-dangers/#Recap_The_Best_Ways_to_Protect_Your_Public_Wi-Fi_Connection | accessed on 24 July 2021.

111 Online article | Business Insider | "'what is Bluetooth?' A beginner's guide to the wireless technology" | 20 May 2020 | https://www.businessinsider.com/what-is-bluetooth | accessed on 25 August 2021.

112 Website | OneTemp | "Bluetooth Low Energy: A Closer Look" | https://www.onetemp.com.au/bluetooth-low-energy-a-closer-look | accessed on 25 August 2021.

113 Cybsersecurity Information Sheet | NSA | "Securing Wireless Devices in Public Settings" | July 2021 | https://media.defense.gov/2021/Jul/29/2002815141/-1/-1/0/CSI_SECURING_WIRELESS_DEVICES_IN_PUBLIC.PDF | accessed on 25 August 2021.

114 Blog post | AT&T Business | "Bluetooth Security Risks Explained" | 11 June 2020 | https://cybersecurity.att.com/blogs/security-essentials/bluetooth-security-risks-explained | accessed on 22 July 2021.

115 Blog post | Bluetooth Blog | "Proximity and RSSI" | 21 September 2015 | https://www.bluetooth.com/blog/proximity-and-rssi/ | accessed on 23 July 2021.

116 Online article | Info Security | "From BIAS to Sweyntooth: Eight Bluetooth Threats to Network Security" | 21 December 2020 | https://www.infosecurity-magazine.com/opinions/bluetooth-threats-network/ | accessed on 2 July 2021.

117 Online publication | Franco Zappa | "BIAS: Bluetooth Impersonation Attacks" | https://francozappa.github.io/about-bias/publication/antonioli-20-bias/antonioli-20-bias.pdf | accessed on 5 July 2021.

[118] Online article | Threat Post | "Bluetooth Bugs Allow Impersonation Attacks on Legions of Devices" | 19 May 2020 | https://threatpost.com/bluetooth-bugs-impersonation-devices/155886/ | accessed on 2 July 2021.

[119] Online article | InformaTech | "Bluetooth Security Weaknesses Pile Up, While Patching Remains Problematic" | 24 September 2020 | https://www.darkreading.com/endpoint/bluetooth-security-weaknesses-pile-up-while-patching-remains-problematic/d/d-id/1339009 | accessed on 2 July 2021.

[120] Website | Armis.com | "Bleedingbit" | 2021 | https://www.armis.com/bleedingbit/ | accessed on 5 July 2021.

[121] Online article | TechCrunch | "Fetch Robotics' CEO on the Company's Acquisition and the Future of Warehouse Robots" | https://techcrunch.com/2018/11/01/bleedingbit-security-flaws-bluetooth-wireless-networks/ | accessed on 2 July 2021.

[122] Website | Armis.com | "BlueBorne" | 2021 | https://www.armis.com/blueborne/ | accessed on 2 July 2021.

[123] Online article | Threat Post | "Wireless 'BlueBorne' Attacks Target Billions of Bluetooth Devices" | 12 September 2017 | https://threatpost.com/wireless-blueborne-attacks-target-billions-of-bluetooth-devices/127921/ | accessed on 2 July 2021.

[124] Online article | ZDNet | "Two Billion Devices Still Vulnerable to Blueborne Flaws a Year after Discovery" | 13 September 2018 | https://www.zdnet.com/article/two-billion-devices-still-exposed-after-blueborne-vulnerabilities-reveal/ | accessed on 2 July 2021.

[125] Online article | Vilabin | "Bluetooth Vulnerabilities: Bluetooth Threats to Network Security" | 21 December 2020 | https://vilabin.com/article/bluetooth-vulnerabilities-bluetooth-threats-network-security/ | accessed on 5 July 2021.

[126] Online article | The Cyber Security Place | "Bluetooth, We Have a Problem" | 25 September 2020 | https://thecybersecurityplace.com/bluetooth-problem/ | accessed on 2 July 2021.

[127] Online article | Threatpost | "Apple's 'Find My' Network Exploited via Bluetooth" | 13 May 2021 | https://threatpost.com/apple-find-my-exploited-bluetooth/166121/ | accessed on 23 July 2021.

[128] Blog post | Trail of Bits | "You could have Invented that Bluetooth Attack" | 1 August 2018 | https://blog.trailofbits.com/2018/08/01/bluetooth-invalid-curve-points/ | accessed on 2 July 2021.

[129] Website | Knobattack.com | "Key Negotiation of Bluetooth Attack: Breaking Bluetooth Security" | 2019 | https://knobattack.com/ | accessed on 5 July 2021.

[130] Online article | Security Boulevard | "New Bluetooth Vulnerability, KNOB Attack Can Manipulate the Data Transferred Between Two Paired Devices" | 20 August 2019 | https://securityboulevard.com/2019/08/new-bluetooth-vulnerability-knob-attack-can-manipulate-the-data-transferred-between-two-paired-devices/ | accessed on 2 July 2021.

[131] Online article | Vilabin | "Bluetooth Vulnerabilities: Bluetooth Threats to Network Security" | 21 December 2020 | https://vilabin.com/article/bluetooth-vulnerabilities-bluetooth-threats-network-security/ | accessed on 5 July 2021.

[132] Online article | Asset Research Group | "Unleashing Mayhem Over Bluetooth Low Energy" | 14 July 2020 | https://asset-group.github.io/disclosures/sweyntooth/ | accessed on 2 July 2021.

[133] Online article | HealthITSecurity | "FDA Warns Medical Device Bluetooth Security Flaw Could Disrupt Function" | 5 March 2020 | https://healthitsecurity.com/news/fda-warns-medical-device-bluetooth-security-flaw-could-disrupt-function | accessed on 2 July 2021.

[134] Online article | Malwarebytes Labs | "Bluetooth Beacons: One Free Privacy Debate with Your Next Order" | 30 June 2020 | https://blog.malwarebytes.com/privacy-2/2020/06/bluetooth-beacons-one-free-privacy-debate-with-your-next-order/ | accessed on 23 July 2021.

[135] Blog post | Blennd | "What is Google Beacon: How Google Beacon Project Affects SEO" | 6 September 2018 | https://blennd.com/what-is-google-beacon-technology-seo/ | accessed on 2 July 2021.

[136] Website | Physicalweb.com | https://google.github.io/physical-web/ | accessed on 5 July 2021.

[137] Online article | Pew Research Center | "More Americans Using Smartphones for Getting Directions, Streaming TV" | 29 January 2016 | https://www.pewresearch.org/fact-tank/2016/01/29/us-smartphone-use/ | accessed on 2 July 2021.

[138] Online article | Google | "Helping Public Health Officials Combat COVID-19" | 3 April 2020 | https://www.blog.google/technology/health/covid-19-community-mobility-reports/ | accessed on 2 July 2021.

[139] Online article | Reuters | "In Coronavirus Fight, Oft-Criticized Facebook Data Aids U.S. Cities, States" | 2 April 2020 | https://www.reuters.com/article/health-coronavirus-facebook-location/in-coronavirus-fight-oft-criticized-facebook-data-aids-u-s-cities-states-idUSKBN21K3BJ | accessed on 2 July 2021.

[140] Blog post | Google: The Keyword | "Helping Public Health Officials Combat COVID-19" | 3 April 2020 | https://www.blog.google/technology/health/covid-19-community-mobility-reports/ | accessed on 5 July 2021.

[141] Website | Nearfieldcommunication.org | "How NFC Works | 2017 | http://nearfieldcommunication.org/how-it-works.html | accessed on 24 July 2021.

[142] Website | Nearfieldcommunication.org | "Security Concerns with NFC Technology" | 2017 | http://nearfieldcommunication.org/nfc-security.html | accessed on 24 July 2021.

[143] Online article | Help Net Security | "Protect Your Smartphone from Radio-Based Attacks" | 19 July 2021 | https://www.helpnetsecurity.com/2021/07/19/smartphone-radio-based-attacks/ | accessed on 24 July 2021.

[144] Online article | Ghostery | "Online Personas and What Not to Share" | 17 March 2020 | https://www.ghostery.com/online-personas-and-what-not-to-share/ | accessed on 2 July 2021.

[145] Online article | Ghostery | "Simple Steps to Manage Your Digital Footprint" | 10 March 2020 | https://www.ghostery.com/easy-ways-to-manage-your-digital-footprint/ | accessed on 4 July 2021.

[146] Online article | Ghostery | "Improve Your Privacy on Social Media" | 5 May 2020 | https://www.ghostery.com/improve-your-privacy-on-social-media/ | accessed on 26 July 2021.

[147] Online article | "Is Your Social Media Activity Leaving You Open to Cyber Attacks?" | 31 August 2020 | https://socialmediaexplorer.com/content-sections/tools-and-tips/is-your-social-media-activity-leaving-you-open-to-cyber-attacks/ | accessed on 30 July 2021.

[148] Online article | Ghostery | "The Privacy Paradox: Concerns vs Behaviors" | 24 March 2020 | https://www.ghostery.com/the-privacy-paradox-concerns-vs-behavior/ | accessed on 26 July 2021.

[149] Online article | Experian | "How to Manage Your Privacy Settings on Social Media" | 30 March 2018 | https://www.experian.com/blogs/ask-experian/how-to-manage-your-privacy-settings-on-social-media/ | accessed on 4 July 2021.

150 Online publication | The Wall Street Journal | "What Hackers Can Learn About You From Your Social-Media Profile" | 8 June 2021 | https://www-wsj-com.cdn.ampproject.org/c/s/www.wsj.com/amp/articles/what-hackers-can-learn-about-you-from-your-social-media-profile-11623157200 | accessed on 23 July 2021.

[151] Online article | Lifewire | https://www.lifewire.com/what-is-facebook-3486391 | accessed on 4 October 2021

[152] Online article | Social Media Today | "Facebook Outlines Its Ad Review Process to Provide More Transparency on Its System" | 20 May 2021 | https://www.socialmediatoday.com/news/facebook-outlines-its-ad-review-process-to-provide-more-transparency-on-its/600580/ | accessed on 4 July 2021.

[153] Website | Facebook.com | "How Ads Work on Facebook" | 2021 | https://www.facebook.com/help/516147308587266/how-ads-work-on-facebook/?helpref=hc_fnav | accessed on 4 July 2021.

[154] Online article | Social Media Today | "Facebook Outlines New Differential Privacy Framework to Protect User Information in Shared Datasets" | 3 June 2020 | https://www.socialmediatoday.com/news/facebook-outlines-new-differential-privacy-framework-to-protect-user-inform/579167/ | accessed on 5 August 2021.

[155] Online article | How-to Geek | "7 Important Facebook Privacy Settings to Change Right Now" | 2 June 2021 | https://www.howtogeek.com/727135/7-important-facebook-privacy-settings-to-change-right-now/ | accessed on 5 August 2021.

[156] Online article | Forbes | "Facebook Cites Russia and Iran as Its Top Sources Of Disinformation [Infographic]" | 27 May 2021 | https://www.forbes.com/sites/niallmccarthy/2021/05/27/facebook-cites-russia-and-iran-as-its-top-sources-of-disinformation-infographic/ | accessed on August 2021.

[157] Online article | How-toGeek | "6 Things You Should Never Share on Facebook and Social Media" |15 May 2021 | https://www.howtogeek.com/723834/6-things-you-should-never-share-on-facebook-and-social-media/ | accessed on 3 August 2021.

[158] Online article | Ghostery | "How to Change Your Privacy Settings on Facebook and Instagram" | 12 May 2020 | https://www.ghostery.com/how-to-change-your-privacy-settings-on-facebook-and-instagram/ | accessed on 5 August 2021.

[159] Online article | Forbes | "All The Ways Facebook Tracks You And How To Stop It" | 8 May 2021 | https://www-forbes-com.cdn.ampproject.org/c/s/www.forbes.com/sites/kateoflahertyuk/2021/05/08/all-the-ways-facebook-tracks-you-and-how-to-stop-it/amp/ | accessed on 3 August 2021.

[160] Online article | Forbes | All The Ways Facebook Tracks You And How To Stop It" | 8 May 2021 | https://www-forbes-com.cdn.ampproject.org/c/s/www.forbes.com/sites/kateoflahertyuk/2021/05/08/all-the-ways-facebook-tracks-you-and-how-to-stop-it/amp/ | accessed on 3 August 2021.

[161] Online article | HackRead | "Facebook Exposed User Data to Thousands of App Developers" | 2 July 2020 | https://www.hackread.com/facebook-exposed-user-data-to-app-developers/ | accessed on 3 August 2021.

[162] Online article | How-to Geek | "How to Restrict Someone on Facebook" | 29 May 2021 | https://www.howtogeek.com/728204/how-to-restrict-someone-on-facebook/ | accessed on 5 August 2021.

[163] Online article | Webopedia | https://www.webopedia.com/definitions/facebook-messenger/| accessed on 4 October 2021

[164] Online article | Makeuseof | https://www.makeuseof.com/tag/what-is-instagram-how-does-instagram-work/ | accessed on 4 October 2021

165 Online article | Ghostery | "How to Change Your Privacy Settings on Facebook and Instagram" | 12 May 2020 | https://www.ghostery.com/how-to-change-your-privacy-settings-on-facebook-and-instagram/ | accessed on 9 August 2021.

[166] Online article | Wired | "How to Stop Instagram from Tracking Everything You Do" | 6 June 2020 | https://www.wired.co.uk/article/instagram-story-ads-privacy-delete | accessed on 9 August 2021.

[167] Online article | How-toGeek | "How to Turn off Message Requests in Instagram" | 2 May 2021 | https://www.howtogeek.com/721386/how-to-turn-off-message-requests-in-instagram/ | accessed on 9 August 2021.

[168] Website | Facebook.com | "How Can I Adjust How Ads on Instagram are Shown to me Based on Data About my Activity from Partners?" | 2021 | https://www.facebook.com/help/instagram/2885653514995517?helpref=faq_content | accessed on 9 August 2021.

[169] Online article | LinkedIn Help | https://www.linkedin.com/help/linkedin/answer/111663/what-is-linkedin-and-how-can-i-use-it-?lang=en | accessed on 4 October 2021

170 Online article | Security Intelligence | "Social Engineering: How to Keep Security Researchers Safe" | 16 May 2021 | https://securityintelligence.com/articles/social-engineering-keep-security-researchers-safe/ | accessed on 22 July 2021.

[171] Online article | CQCore | "Are You Linked In?" | 24 March 2021 | https://www.cqcore.uk/are-you-linked-in/ | accessed on 4 July 2021.

[172] Website | Support.snapchat.com | "Snapchat Support" | 2021 | https://support.snapchat.com/en-US | accessed on 9 August 2021.

[173] Online article | Common Sense Media | "Parents' Ultimate Guide to Snapchat" | 9 March 2021 | https://www.commonsensemedia.org/blog/parents-ultimate-guide-to-snapchat | accessed on 9 August 2021.

[174] Website | Snap.com | "Control Over Your Information" | 2021 | https://snap.com/en-US/privacy/privacy-policy/#control-over-your-information | accessed on 9 August 2021.

[175] Blog post | Privacycrypts | "Snapchat Privacy Settings And Tips In 2021" | 31 December 2020 | https://privacycrypts.com/blog/snapchat-privacy-settings-tips/ | accessed on 9 August 2021.

[176] Website | Snap.com | "How We Use Your Information" | 2021 | https://snap.com/en-US/privacy/your-information | accessed on 9 August 2021.

[177] Online article | Influencer Marketing Hub | https://influencermarketinghub.com/what-is-tiktok/ | accessed on 4 October 2021

[178] Online article | 9TO5Mac | "TikTok Privacy Policy Now Says App Will Collect 'Faceprints and Voiceprints'" | 4 June 2021 | https://9to5mac.com/2021/06/04/tiktok-privacy-policy-now-says-app-will-collect-faceprints-and-voiceprints/ | accessed on 9 August 2021.

[179] Website | Lifewire | https://www.lifewire.com/what-exactly-is-twitter-2483331 | accessed on 1 October 2021

[180] Online article | Twitter Help Center | https://help.twitter.com/en/resources/new-user-faq | accessed on 1 October 2021

[181] Website | Help.twitter.com | "What Can we Help you Find?" | 2021 | https://help.twitter.com/ | accessed on 11 August 2021.

[182] Online article | Twitter Help Center | https://help.twitter.com/en/using-twitter/following-faqs.html | accessed on 1 October 2021

[183] Online article | Twitter Help Center | "How to Control your Twitter Experience" | 2021 | https://help.twitter.com/en/safety-and-security/control-your-twitter-experience | accessed on 11 August 2021.

[184] Online article | Twitter Help Center | "How to Protect your Personal Information" | 2021 | https://help.twitter.com/en/safety-and-security/twitter-privacy-settings | accessed on 11 August 2021.

[185] Online article | Malwarebytes Labs | "Have I Been Pwned?"– What Is It and What To Do When You *Are* Pwned" | 20 May 2021 | https://blog.malwarebytes.com/awareness/2021/05/have-i-been-pwnd-what-is-it-and-what-to-do-when-you-are-pwned/ | accessed on 23 July 2021.

[186] Online article | Twitter Help Center | "How to Protect your Personal Information" | 2021 | https://help.twitter.com/en/safety-and-security/twitter-privacy-settings | accessed on 11 August 2021

[187] Online article | Twitter Help Center | "How to Protect and Unprotect your Tweets" | 2021 | https://help.twitter.com/en/safety-and-security/how-to-make-twitter-private-and-public | accessed on 11 August 2021.

[188] Online article | Twitter Help Center | "About Account Security" | 2021 | https://help.twitter.com/en/safety-and-security/account-security-tips | accessed on 11 August 2021.

[189] Online article | Ghostery | "How to Change Your Privacy Settings on Twitter and YouTube" | 19 May 2020 | https://www.ghostery.com/how-to-change-your-privacy-settings-on-twitter-and-youtube/ | accessed on 11 August 2021.

[190] Online article | Twitter Help Center | "How to use two-factor authentication" | 2021 | https://help.twitter.com/en/managing-your-account/two-factor-authentication | accessed on 11 August 2021.

[191] Online article | Bitdefender | "Despite All the Advice, 97.7% of Twitter Users Have Still Not Enabled Two-Factor Authentication" | 27 July 2021 | https://www.bitdefender.com/blog/hotforsecurity/despite-all-the-advice-97-7-of-twitter-users-have-still-not-enabled-two-factor-authentication | accessed on 11 August 2021.

[192] Online article | Lifewire | https://www.lifewire.com/youtube-101-3481847 | accessed on 4 October 2021

[193] Website | Support.google.com | "Change your Subscription Privacy Settings" | 2021 | https://support.google.com/youtube/answer/7280190?hl=en&ref_topic=9257519 | accessed on 10 August 2021.

[194] Website | Youtube.com | "Privacy Controls" | 2021 | https://www.youtube.com/howyoutubeworks/user-settings/privacy/ | accessed on 10 August 2021.

[195] Online article | Ghostery | "How to Change Your Privacy Settings on Twitter and YouTube" | 19 May 2020 | https://www.ghostery.com/how-to-change-your-privacy-settings-on-twitter-and-youtube/ | accessed on 10 August 2021.

[196] Website | Support.google.com | "Update your Location Settings in YouTube Music" | 2021 | https://support.google.com/youtubemusic/answer/9015821 | accessed on 10 August 2021.

[197] Website | Support.google.com | "Watching "Made for Kids" Content" | 2021 | https://support.google.com/youtube/answer/9632097 | accessed on 10 August 2021.

[198] Online article | AndroidCentral | "How to Set Up Supervised Accounts for your Kids on YouTube" | 8 May 2021 | https://www.androidcentral.com/how-set-supervised-accounts-your-kids-youtube | accessed on 10 August 2021.

[199] Online article | Techopedia | https://www.techopedia.com/definition/5359/google | accessed on 4 October 2021

[200] Online article | Axios | "What Google Knows About You" | 11 March 2019 | https://www.axios.com/what-google-knows-about-you-3f6c9b20-4406-4bda-8344-d324f1ee0816.html | accessed on 4 July 2021.

[201] Lifewire | https://www.lifewire.com/how-to-stop-google-from-tracking-you-4175369 | accessed on 10 September 2019

[202] Online article | Visual Capitalist | "What Does Google Know About You?" | 10 August 2018 | https://www.visualcapitalist.com/what-does-google-know-about-you/ | accessed on 4 July 2021.

[203] Online article | Wired | "All the Ways Google Tracks You—And How to Stop It" | 27 May 2019 | https://www.wired.com/story/google-tracks-you-privacy/ | accessed on 4 July 2021.

[204] ZDNet | https://www.zdnet.com/article/google-revamps-privacy-policy-to-give-users-more-controls-over-assistant-voice-recordings/

[205] Blog post | Google: The Keyword | Doing More to Protect your Privacy with the Assistant" | 23 September 2019 | https://www.blog.google/products/assistant/doing-more-protect-your-privacy-assistant/ | accessed on 4 July 2021.

[206] Online article | Forbes | "Google to Fix Malicious Invites Issue for 1 Billion Calendar Users" | 9 September 2019 | https://www.forbes.com/sites/daveywinder/2019/09/09/google-finally-confirms-security-problem-for-15-billion-gmail-and-calendar-users/#e7f62c3279fa | accessed on 4 July 2021.

[207] Online article | Black Hills Information Security | "Google Calendar Event Injection with MailSniper" | 1 November 2017 | https://blackhillsinfosec.com/google-calendar-event-injection-mailsniper/ | accessed on 4 July 2021.

[208] Online article | Fast Company | https://www.fastcompany.com/company/amazon | accessed on 4 October 2021

[209] Website | Amazon | "How Amazon Collects Your Personal Information" | https://www.amazon.com/gp/help/customer/display.html/?nodeId=GSXETHUPY4UM7CRD | accessed on 25 August 2021.

[210] Website | Amazon | "How Amazon Uses Your Personal Information" | https://www.amazon.com/gp/help/customer/display.html/ref=hp_left_v4_sib?ie=UTF8&nodeId=G6CVQVUVGMD3BJQ2 | accessed on 25 August 2021.

[211] Website | Amazon | "Internet-Based Ads" | https://www.amazon.com/gp/help/customer/display.html/?nodeId=GLVB9XDF9M8MU7UZ | accessed on 25 August 2021.

[212] Website | Amazon | "Amazon.com Privacy Notice" | 12 February 2021 | https://www.amazon.com/gp/help/customer/display.html?nodeId=GX7NJQ4ZB8MHFRNJ | accessed on 25 August 2021.

[213] Online article | Wired | "All the Ways Amazon Tracks You—and How to Stop It" | 22 June 2021 | https://www.wired.com/story/amazon-tracking-how-to-stop-it/ | accessed on 9 August 2021.

[214] Online article | The Hacker News | "Your Amazon Devices to Automatically Share Your Wi-Fi With Neighbors" | 31 May 2021 | https://thehackernews.com/2021/05/your-amazon-devices-to-automatically.html?utm_source=feedburner&utm_medium=feed&utm_campaign=Feed%3A+TheHackersNews+%28The+Hackers+News+-+Cyber+Security+Blog%29 | accessed on 9 August 2021.

[215] Website | Amazon | "About Security Alerts" | https://www.amazon.com/gp/help/customer/display.html/ref=hp_left_v4_sib?ie=UTF8&nodeId=GLXNK37D6R3WGXKW | accessed on 25 August 2021.

[216] Website | Amazon | "Browser Extensions & Privacy" | https://www.amazon.com/gp/help/customer/display.html?nodeId=G8V457F4P763VW8D | 25 August 2021.

[217] Online article | C|Net | "8 Ways to Protect Your Amazon Echo Privacy while Working from Home" | 26 March 2020 | https://www.cnet.com/home/smart-home/8-ways-to-protect-your-amazon-echo-privacy-while-working-from-home/ | accessed on 9 August 2021.

[218] Online article | AndroidCentral | "What You Need to Know About Amazon Sidewalk Before You Decide to Opt Out" | 26 June 2021 | https://www.androidcentral.com/what-you-need-know-about-amazon-sidewalk-you-decide-opt-out | accessed on 9 August 2021.

[219] Online article | The Verge | "How to Opt Out of (or into) Amazon's Sidewalk Network" | 1 June 2021 | https://www.theverge.com/22463257/amazon-sidewalk-privacy-how-to-opt-out | accessed on 9 August 2021.

[220] Website | SanctionScanner | https://sanctionscanner.com/knowledge-base/money-service-business-133 |accessed on 1 October 2021

[221] Website | Comply Advantage | https://complyadvantage.com/knowledgebase/anti-money-laundering/money-services-business/ | accessed on 1 October 2021

[222] Website | PayPal | https://www.paypal.com/myaccount/privacy/privacyhub | accessed on 2 September 2021.

[223] Website | Finextra | https://www.finextra.com/newsarticle/38008/twitter-tip-jar-users-cite-paypal-privacy-concerns-hours-after-rollout | accessed on 8 June 2021

[224] Website | BleepingComputer – Ax Sharma | https://www.bleepingcomputer.com/news/security/twitter-tip-jar-may-expose-paypal-address-sparks-privacy-concerns/ | accessed on 8 June 2021

[225] Website | Venmo | https://venmo.com/about/us/ | accessed on 8 June 2021

[226] Online Article | EFF | "Venmo Takes Another Step Toward Privacy" | 21 July 2021 | https://www.eff.org/deeplinks/2021/07/venmo-takes-another-step-toward-privacy | accessed on 2 September 2021.

[227] Website | PayPal | https://www.paypal.com/webapps/mpp/ua/privacy-full#dataCollect | accessed on 8 June 2021

[228] Website | Venmo | https://help.venmo.com/hc/en-us/articles/210413717-Payment-Activity-Privacy | accessed on 8 June 2021

[229] Online article | Android Central | "How to remove location data from photos on Android" | 3 June 2020 | https://www.androidcentral.com/how-remove-location-data-photos-android | accessed on 8 August 2021.

[230] Blog post | Backlight | "Google Photos and privacy: How to keep your photos safe" | https://backlightblog.com/google-photos-privacy | accessed on 1 October 2021

[231] Online article | How-toGeek | "How to Hide People from Memories in Google Photos" | 21 May 2021 | https://www.howtogeek.com/729788/how-to-hide-people-from-memories-in-google-photos/ | accessed on 11 August 2021.

[232] Google | Website | "Watch & Manage Your Memories" | https://support.google.com/photos/answer/9454489?hl=en&co=GENIE.Platform%3DAndroid#zippy=%2Ccontrol-which-memories-you-receive%2Chide-someone | accessed on 1 October 2021

www.ingramcontent.com/pod-product-compliance
Lightning Source LLC
Chambersburg PA
CBHW051756200326
41597CB00025B/4579